Springer Series in Wireless Technology

Series editor

Ramjee Prasad, Aalborg, Denmark

Springer Series in Wireless Technology explores the cutting edge of mobile telecommunications technologies. The series includes monographs and review volumes as well as textbooks for advanced and graduate students. The books in the series will be of interest also to professionals working in the telecommunications and computing industries. Under the guidance of its editor, Professor Ramjee Prasad of the Center for TeleInFrastruktur (CTIF), Aalborg University, the series will publish books of the highest quality and topical interest in wireless communications.

More information about this series at http://www.springer.com/series/14020

Punnarumol Temdee · Ramjee Prasad

Context-Aware Communication and Computing: Applications for Smart Environment

Springer

Punnarumol Temdee
School of Information Technology
Mae Fah Luang University
Chiang Rai
Thailand

Ramjee Prasad
CTIF Global Capsule and Future
Technologies for Business Ecosystem
Innovation (FT4BI)
Aarhus University
Herning
Denmark

ISSN 2365-4139 ISSN 2365-4147 (electronic)
Springer Series in Wireless Technology
ISBN 978-3-319-59034-9 ISBN 978-3-319-59035-6 (eBook)
DOI 10.1007/978-3-319-59035-6

Library of Congress Control Number: 2017941471

Printed on acid-free paper

This Springer imprint is published by Springer Nature
The registered company is Springer International Publishing AG
The registered company address is: Gewerbestrasse 11, 6330 Cham, Switzerland

इन्द्रियाणि पराण्याहुरिन्द्रियेभ्यः परं मनः |
मनसस्तु परा बुध्दियोर बुद्धेः परतस्तु सः ||

Transliteration

Indriyani Paraanyahurindriyebhyah param manah |

Manasastu paraa buddhiyor buddheh partastu sah ||

Explanation

It is affirmed that observance (senses) makes us superior, but more than observance, awareness (mind) is superior, but more than being aware, the perseverance is superior, and more than perseverance that which is superior is the individual perspicacity.

|| 3.42 ||

Preface

This book introduces the concept of context-aware computing and its applications in various areas. It is designed particularly for the beginners who would like to design and develop the smart environment with context-aware computing. The user-friendly content is offered not only for the readers from Information Communication Technology related areas but also other professional domains such as Humanity, Public Health, Social Science, etc. The foundation of context-aware computing is described in this book such as definitions, categories, characteristics, context awareness, etc. Here, the elements of context-aware applications including context acquisition, context modeling, context reasoning, context distribution, and context adaptation are also emphasized. Communication and security are introduced so that the readers understand how all components work together with the security awareness. Additionally, some existing middleware and applications are presented so that the readers get the idea for selecting the right tool for their requirements and developing their applications appropriately. More importantly, the author's perspectives accordingly to context definitions, its awareness, and future context applications are suggested in this book. The ultimate goal of this book is to expand the contribution of context-aware computing to new professional areas where the utilization of personalized and rationalized applications as smart environments are required.

Acknowledgements

We would like to express our gratitude to many people who provided supports for accomplishing this book. We also wish to thank all contributions from existing works in context-aware computing and other related areas. Finally, our sincere gratitude goes to all personal supports from family, friends, students, and colleagues.

Chiang Rai, Thailand — Punnarumol Temdee
Herning, Denmark — Ramjee Prasad

Contents

Chapter 1
Introduction to Context-Aware Computing

Abstract This chapter aims to introduce the concept and foundation of context-aware computing. Variety kinds of services and applications of context-aware computing can be seen everywhere with a wide range of application domains such as Information Communication Technology (ICT), Health Science, Humanity, Social Science, etc. Most applications and services of context-aware computing are not only in the imaginary any longer. Many innovations in context-aware computing are now under the development to achieve the ultimate goal which is to support the convenient use for the users.

Context-aware computing is widely used in many applications nowadays varying from desktop applications, web applications, mobile applications, to the Internet of Things (IoT). We can simply see many context-aware applications embedded into our daily life at our home or our cars to those in the office, the factory, the hospital, the airport, etc. For example, your home sends the greeting voice in the morning when you wake up and serve you with your favorite coffee and the morning news. In your office, the lights are turned on and off appropriately when you walk pass through them. In the shopping mall, your refrigerator sends you the reminding message about what you have to buy. In your mobile phones, you can be suggested to change your lifestyle if you have eaten too much sweet this week. At the hospital, the nurse comes to you when you are about to fall in your room. Most applications and services of context-aware computing are not only in the imaginary any longer. Many innovations in context-aware computing are now under the development to achieve the ultimate goal which is to support the appropriate response to the user and the environment. This chapter aims to introduce the concept, foundation, and contribution of context-aware computing not only in the Information Communication Technology (ICT) related areas but also other application domains so that the context-aware computing will be applied succesfully to other application domains.

P. Temdee and R. Prasad, *Context-Aware Communication and Computing: Applications for Smart Environment*, Springer Series in Wireless Technology,
DOI 10.1007/978-3-319-59035-6_1

1.1 Context of Context-Aware Computing

Context-aware applications can adapt their functions, contents, and interfaces according to the user's current situation with less distraction of the users. More specifically, such requests can discover the contextual information such as locations, networks, nearby persons or objects, etc. Context-aware computing can be introduced with some related visions such as ubiquitous/pervasive computing, invisible computing, proactive computing, ambient intelligence, sentient computing, etc. (Loke 2006). Many pieces of literature have demonstrated the overlapping and relating visions among them.

The term context-aware computing is widely introduced after the introduction of the article entitled "The computer of the 21st Century" (Weiser 1991). The computing world in the future is predicted that it will consist of small, seamlessly interconnected computing devices in which the users can wear some of them. Ubiquitous computing can thus also be called synonymously as pervasive computing. Pervasive computing frequently refers to the vision of pervading devices or computers. Weiser (Weiser 1991) has also stated that context-awareness is an essential building block for realizing the vision of ubiquitous computing. One possible reason is that the user's context is normally used to execute the application unconsciously. Ubiquitous computing is claimed as the third wave of computing, which the computers is blended into our everyday lives inconspicuously. Therefore, the essential characteristic of ubiquitous computing is not only anywhere and anytime but also context-awareness service. Since then, many researchers have been studied the research topics surrounding both ubiquitous computing and context-awareness.

Invisible computing (Norman 1998; Borriello 2000) focuses on using the computer for performing the tasks rather than using it as the tools. It shares the same idea with context-aware computing that there should be the least destruction to the users. Proactive computing (Tennenhouse 2000) identifies what the user requires and uses it for taking the action on user's behalf so that the user can focus on higher level tasks rather than the user interfaces. This vision shares the idea of context-aware computing that the unconscious interaction among the users and the computing devices are mainly focused. Ambient intelligence (Aarts 2004) focuses on provinding unobtrusive and invisible services in everyday objects for the users. The context-aware computing shares the common idea by involving the employment of user's context for executing those objects to provide the appropriate services rather than acquiring the input from the users explicitly. Sentient computing (Hopper 2000) refers to the systems using sensor and status data to communicate with the users and the applications. This vision shares the same idea of context-aware computing that the system can model the world from sensory information and uses them to execute the appropriate applications. Currently, the Internet of Thing (IoT) is implemented worldwide. It allows the machine to talk to the machine conveniently. The information which is called as the context is collected through different kinds of sensors autonomously (Perera et al. 2014). The context-aware computing plays the primary role in supporting IoT by gathering, manipulating, and delivering the context appropriately accordingly to the requirments of the users and the environment. Since context-aware computing shares similar ideas and involves many different computing visions, the applications of context-aware computing can be broadly found. More detail will be discussed later throughout this book.

1.2 Pathway of Context-Aware Computing

The introduction of context-aware computing can be described as different related research areas. The early research works mainly focus on the definitions of context and its awareness. The first definitions of context depend on the information necessary for different applications (Schilit et al. 1994; Pascoe et al. 1999; Dey 1998) such as environment, location, identity, emotion status, etc. These various types of data are so-called contexts. Once the general definition of context is indeed required, more general or conceptual definitions are introduced. The most cited one is to refer the context as "any information that can be used to characterize the situation of an entity" (Dey and Abowd 2000b). Although many works seem to agree with this general definition, the characterization of context and its awareness remain challenging because of variety kinds of emerged context types in current and future applications.

At the same time, many works have focused on the applications and the services of context-aware computing rather than its definitions. Many context-aware applications and services have been developed to demonstrate and validate the usability and the flexibility according to the context and its status. The early works proposed the useful applications and services without defining the exact definition of context-aware. For example, the "Active Badge Location System" (Want et al. 1992) can be considered as one of the pioneer systems for identifying the location of people in an office environment. For this system, the location information is transmitted through a network of sensors to the central location service. More specifically, this system aims to forward the calls to a phone closest to the user according to the user's location. At the same time, the location-aware tour guides are also one of the popular applications. They provide tourist information according to the user's current locations (Abowd et al. 1997; Fels et al. 1998; Cheverst et al. 2000). Some additional features are added for supporting the convenient trip for the users such as map, navigator, etc.

Although many context-aware systems and services have existed in the last decades, most of them are still facing several limitations. The significant limitations include the difficulty of implementation of relevant complicated methods for capturing, representing and processing the context as well as the methods for adaptation. The main reason is that the system and the service are designed to satisfy the predefined set of contexts without the flexibility of emerging context information. To achieve the effectiveness, flexibility, and scalability of the context-aware applications, the research attention not only focuses on more generalization of context definition but also on generic frameworks supporting variety kinds of context and its awareness. Consequently, many works have been proposing the general frameworks for context-aware systems (Budzik and Hammond 2000; Finkelstein and Savigni 2001; Dey and Abowd 2000a; Hofer et al. 2003) so that the framework can fit any applications appropriately.

The generic framework for the context-aware computing system is designed and developed to work effectively with any context information. The works in this research area cover from non-flexible context model workable with particular applications (Chen and Kotz 2000; Chen et al. 2003; Korpipää et al. 2003) to general flexible and extensible context models (Gu et al. 2004; Fahy and Clarke 2004; Sheng et al. 2004; Ejigu et al. 2007). The generic framework commonly includes tools and methods achieving effectiveness on various aspects causing some relevant research area consequently. Context sensing or acquisition is one of the examples. It involves sensors technology and sensor networking to ensure the complete acquisition of raw context information. After having the raw context information, the processes of modeling, representing and storing of context information are required (Baldauf et al. 2007) to make raw context information to be kept, represented and retrieved appropriately for the further process. Several methods for context modeling frequently rely on the data structure in the context-aware applications (Strang and Linnhoff-Popien 2004) such as graphical model, logic-based model, object-oriented model, ontology-based model, etc. As mentioned before, the context-aware applications must primarily provide the ability to store, maintain and query historical context data. Nowadays, knowledge discovery of historical context data is even gaining much interest due to the big data era (Manyika et al. 2011). The historical context data is important for determining the changes in context patterns and predicting the future context values with many data mining techniques and knowledge discovery methods. This evidence gives the great opportunity for the developers and the researcher to explore the area of the context-aware applications with big data perspective.

Much attention is also paid in processing, aggregating and reasoning of contextual information. It is necessary to have the appropriate process for providing higher level context information from raw sensor data. Data aggregation is the method used for manipulating contextual information with any operation method for constructing higher level context abtraction which is useful for particular applications. Context reasoning represents the process to deduce new or relevant information from the different sources of context data. It is still challenging nowadays how to determine the appropriate processing, aggregating and reasoning methods to satisfy the expectation of the current and the future applications.

Accordingly, to the concept of context awareness, context adaptation is also gaining much interest (Adelstein et al. 2005) because it can make context-aware application able to adapt itself accordingly to the user's preference and situation. The adaptation can also be demonstrated in many different aspects such as the adaptation of functionality, delivered data, user interface, etc. The middleware is usually involved in performing effectively adaptation. Middleware is necessary for context-aware development because it can help developing context-aware application much easier and promoting reusability, extensibility, and scalability of the applications. Additionally, it is also responsible for the integration of design and development of context-aware applications. Currently, there are many kinds of middleware available for the developers and the researchers. Consequently, there are some significant concerns for the selection of appropriate middleware which typically depend heavily on applications.

Security and privacy of context data are the important issues nowadays because the application normally includes sensitive information of people which is required to be protected. The policy to define the ownership and the right access to the systems are needed to be implemented explicitly. Especially for IoT paradigm, security and privacy are even more concerned because the machines now can communicate with each other. More importantly, there is still important argument about the trade-off between security and privacy for IoT paradigm, whereas the applications are expected to provide autonomous and personalized services at the same time.

As mentioned before, it can be clearly seen that the context-aware application is constituted by many research areas. The pathway of the context-aware application can be summarized as shown in Fig. 1.1. Many opportunities are remaining to discover more new findings. This book aims to provide the necessary background knowledge and challenge the readers for pursuing the achievement of this area.

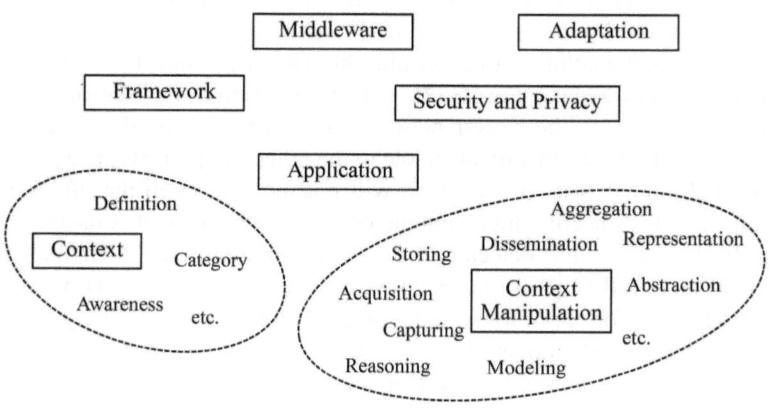

Fig. 1.1 Summarization of Context-Aware Computing Pathway

1.3 Context-Aware Applications

As it can be seen that the context aware applications can be found in many different application domains, this chapter will briefly introduce some important applications gaining intensive attentions worldwide.

1.3.1 Location-Aware Applications

Location-aware applications can be considered as the standard applications for the early stage of context-aware computing because the location is normally defined as the context. Location-aware applications deliver functions or services to the users based on their physical location. For example, the application generates the reminders or provides necessary information to the users as they enter or walk through specific areas. Various technologies can be used for acquiring location contexts such as Global Positioning System (GPS), mobile network, wireless access points, etc. to identify user's entities such as cell phones, laptops, tablet, etc. Then the locations can be chosen to share by the users with location-aware applications. Those applications can provide the users with a variety of contents such as current location on the map, nearby restaurants, notices about traffic condition, etc. At the same time, the applications can also report a user's locations in any social network creating an important marketing opportunity currently. There are many pioneers for location-aware application varying from using only the location to execute the application as Active Badge Location System for call forwarding (Schilit et al. 1993) to using any other involved contexts to perform specific purposes such as tour guide, personal shopping assistant, etc. This section discusses in detail for some applications to briefly show the evolution of location-aware application.

As mentioned before, Active Badge Location System (Want et al. 1992) is well known as a pioneer of location-aware application. It aims to locate the staff in a large organization by using infrared (IR) technology. Active Badge Location System provides the solution to determine individual location by using a tag that can emit a unique code for every 15 s. For this system, the sensors are placed around the building as the sensor network. The server is used for keeping and making the location information available to the other components. The pulse-width modulated IR signal is used because it will not move through the office partition like radio signals. Since it emits a unique code for every 15 s, it is useful in power saving. It also has a light detector in the dark condition. Active Badge Location System has an excellent performance to wear on the breast pocket. However, it has a little dropped performance for wearing on the belt or the waist. Many sensors can be installed and placed on many places such as the exits, entrance, walls, etc. Up to 128 sensors can be connected to the computer at the same time. Four wires sensor networking (2 for power supply, 1 for control/poll, and 1 for data) are used. For the application, the system collects name, location information, and phone network.

With Active Badge Location System, the automatic redirection of incoming calls can be efficiently done. There is less chance of missing some important waiting calls. The system can quickly tell whether a person is in or not. It is easier to ask for a meeting and to identify the visitors in the office space.

The tour guide is also the popular application of location-aware applications (Bellotti et al. 2005; Höpken et al. 2010; Anacleto et al. 2014). After the mobile network has been widely available and affordable, the application is required to be aware of particular contexts such as the current position, the current orientation, and the location history of the users. The large variety kinds of context-aware mobile applications are developed including the well-known one called Cyberguide (Abowd et al. 1997). Cyberguide aims to explore context-aware mobile applications for future computing environments and to develop a knowledgeable handheld tour guide. The primary goal is to prototype a context-aware tour guide based on portable devices by detecting where the tourists are, and what they are looking for, predicting and answering the question that they may pose, and providing interaction with other people and environment. Cyberguide has four main components including Cartographer, Librarian, Navigator, and Messenger. Cartographer is a map component having knowledge of physical surroundings. Librarian is an information component providing access to information that a tourist might encounter. Navigator is used for charting the position of the visitor. Messenger is a wireless communication component supporting Transmission Control Protocol/Internet Protocol (TCP/IP) packets for sending/receiving email. From the screen of portable devices, there are the icons showing user's position and demonstration stations. The users can select the star icon to reveal its name and its information. The users can also search from information pages. The user can be asked to complete a questionnaire. Infrared is used to sense the user's current indoor position because it is low-cost and convenient. Keeping track of last recorded cell location or historical location provides the prediction for user's orientation. For outdoor positioning, a GPS unit is used. Some features are extended to increase interaction with the environment such as the visitors can keep a record of their experiences, the database can be modified by the user, the maps' level details can be varied, and the detail can be automatically chosen.

Personal shopping assistant (Khor 2016) is also one of the popular location-aware applications. Besides location context, other related contexts are employed especially identity context. The mobile phone applications and web services are used as the general architecture for personal shopping assistant (Wu and Natchetoi 2007). The primary purpose of personal shopping assistant is to develop an application that can assist the users to make a smarter shopping plan while saving time and cost. The applications typically can provide essential features necessary for smart shopping. For example, the navigation tool to provide the location of the store to the users, the product information that will be displayed to the user based on search and price comparison, the budget estimation through the shopping list, the current sale and promotion informing when they are close to the shop, etc. Search and Go (SAGO) (Gültekin and Bayat 2014) is a smart location-based mobile shopping application for Android operating system.

This application provides the location information by using Geo-position of mobile device. After the user's location is identified, the product searching of the nearby shops will be performed. The user will obtain the price, sale promotion, stock details, etc., of the products from those nearby shops. The architecture consists of three layers including Resource Layer, Data Access and Extraction Layer, and Presentation Layer (Gültekin and Bayat 2014). The Resource Layer includes all of the related data collected from the local stores. Data Access and Extraction Layer consists of data extraction and tools. Presentation Layer includes sorting of relevant results and displaying the results in a logical and meaningful way to fulfill user's requirements. The numbers of the smart algorithm such as clustering algorithm, greedy search algorithm, etc., are used to ensure the accurate search and result lists. This application can provide the search results in acceptable time duration, but there is no user interaction feature in the application at that point.

1.3.2 Social-Aware Applications

Currently, context-aware computing is adopted by other applications domains where the application that can respond appropriately to the users and the environment is required. With a dramatically growth of social media and advance mobile devices, the applications that are aware of social context has been gaining much interest nowadays (Kabir et al. 2014a, b, c). The notion of social awareness extends the vision of context-aware computing because the applications need to deal with the human who is the social being. Social context is invented as the context that can represent the interaction among people. It is responsible for characterizing multiple users such as the social tie, social group or group dynamics (Schuster et al. 2013; Liang and Cao 2015). More specifically, it can be defined as a set of derived information from direct or indirect interactions among people in both virtual and physical worlds (Liang and Cao 2015; Intayoad et al. 2017). The key requirement of social context-aware applications is the platform or the middleware to support ease development by reducing the complexity of technical works. The challenging of social-aware application is to collect the social context information from various sources, perform modeling and reasoning for the complex situation, mediate or coordinate the variety of social interactions, and manage them in a proper manner.

The early social-aware applications rely on ad-hoc architectures and context representations because the applications are mainly designed to satisfy specific requirements. The social context applications can employ other context information besides social contexts, such as location, time, activity, etc. Later, the separation between acquiring social context information and context management is the important key for application development and maintenance. Consequently, a variety of social context middleware are proposed (Wang et al. 2008). Social context middleware commonly consists of three main components including programming abstraction, system services and cross-layer support (Liang and Cao 2015).

Programming abstraction provides high-level abstraction interfaces for the developers. System services provide the application deployment and execution. Cross-layer support provides the system security, privacy, and quality of service (QoS). Unlike the traditional context-aware middleware, most middleware of social context-aware application presents some challenges especially in supporting multiple users rather than one single user. All among existing middleware for social context application, they have shared some common components and shown some distinct differences. Many social context middleware define different definitions of social contexts. However, they typically share the common points of views of the social context in term of interaction or relationship. For example, Context-Aware Advertising Mediator and Optimizer (CAMEO) (Arnaboldi et al. 2014) defines the social context as "The information that is derived from both virtual and physical social interactions among users." The socially aware and mobile architecture (SAMOA) (Bottazzi et al. 2007) defines the social context as "The information which characterizes the interactions among a group of people who are in physical proximity." The Social Context Information Management System (SCIMS) (Kabir et al. 2012) defines the context as "A set of information that is derived from virtual and physical interactions among users." The Middleware for Managing Mobile Social Ecosystems or Yarta project (Toninelli et al. 2011) defines social as "The information which characterizes the relationships between users who are in physical proximity." Since the social context definitions are defined differently, the detail of architectures and context modeling and reasoning may also be different. For example, CAMEO has distributed architecture and uses object-role model for context modeling and knowledge-based method for context reasoning. SAMOA has distributed architecture and uses ontology-based methods for context modeling and reasoning. SCIMS has centralized architecture and uses ontology-based methods for context modeling and reasoning. Yarta has distributed architecture and uses ontology-based methods for both context modeling and reasoning. Currently, the numbers of social context middleware increase rapidly. Many of them aim to satisfy the users from the different area such as health science, public health, business, etc. However, the literature also shows that the social context-aware applications are expected to promote security and privacy as well as the ease development for non-technical users.

Social-aware applications can be viewed as two different points of views including data-centric and interaction-centric applications (Brézillon et al. 2014). For data-centric applications, social context information is used for executing the application's behavior such as social roles, social situations, social relationships, etc. The user's relationalties, which can be considered as the connection-oriented relationships, are mainly used for application's behvaior. The connection-oriented relationships can be classified as object-centric and people-centric relationships (Kourtellis et al. 2010) respectively. For object-centric relationship, the relationship is identified between people who have something in common such as the interests, activities, groups, etc. The examples are preference inferring (Mislove et al. 2006), resource sharing (Li and Dabek 2006), etc. On the other hand, the people-centric relationship is defined as a formal definition of a direct connection between people.

For example, one person identifies other persons with a particular type of connections such as the close friend, family member, colleague, etc. This type of relationship can be used in many applications such as preference audio turning (Biamino 2011), review quality quantifying (Lu et al. 2010), a socially-aware phone call application (Kabir et al. 2014a, b, c), etc. In general, to develop a data-centric based social-aware application needs two essential requirements. Firstly, applications have to acquire its user's social context information correctly from external sources both directly and being derived from the available context. Secondly, the applications may need to allow their users to share social context information with other users. The security and privacy functions are thus the important concerns.

For interaction-centric applications, the interaction-oriented social relationships among people dominate the applications' behavior such as peer and group relationships, etc. These applications can assist users to enrich their social interactions and enhance their well-being in their daily lives (Modes 2012). There are many interaction-centric applications nowadays. For example, Sociotelematic application (Kabir et al. 2014a, b, c) can enable the safe driving by using the collaborative relationship among the driver. The companies can create the optimized advertisement strategy by identifying the influential individuals (Adams 2011). The coalition scheme can be suggested to the similar buyer by analyzing relation among the group members (Boongasame et al. 2012). Additionally, besides the interaction itself, the combination with other contexts can also empower the usefulness of applications. For example, the location-based groups are very useful for applications of security and public health. For example, the security department can perform crowd detection and criminal analysis (Yu et al. 2012). The health department can monitor the spread of infectious disease and take action on time (Eubank et al. 2004). Moreover, there are some significant concerns for interaction-centric applications. Firstly, the applications should support interactions having agreed relationships. The runtime environment supporting system adaptation is required to facilitate real-time interactions.

1.4 Book Preview

This book mainly introduces the concept of a context-aware computing which can be seen by its applications and the related research areas as shown in this chapter. Chapter. 2 is designed to explain in detail of context, its definition, its characteristics, and its awareness. Chapter. 3 describes elements of context-aware applications. Chapter. 4 accounts for the foundation of communication for context-aware applications. Chapter. 5 introduces the principle concept of security required for context-aware applications. Chapter. 6 shows some examples of context-aware middleware and applications, especially for the smart environments. Finally, the future direction of context-aware applications, particularly for social context and its applications in healthcare and education domains, is included at the end of this book.

References

Aarts, E. (2004). Ambient intelligence: a multimedia perspective. *MultiMedia, IEEE, 11*(1), 12–19.

Abowd, G. D., Atkeson, C. G., Hong, J., Long, S., Kooper, R., & Pinkerton, M. (1997). Cyberguide: a mobile context-aware tour guide. *Wireless Networks, 3*(5), 421–433.

Adams, P. (2011). *Grouped: How small groups of friends are the key to influence on the social web*. New Riders.

Adelstein, F., Gupta, S. K. S., GR III, & Schwiebert, L. (2005). Fundamentals of Mobile and Pervasive Computing.

Anacleto, R., Figueiredo, L., Almeida, A., & Novais, P. (2014). Mobile application to provide personalized sightseeing tours. *Journal of Network and Computer Applications, 41,* 56–64.

Arnaboldi, V., Conti, M., & Delmastro, F. (2014). CAMEO: a novel context-aware middleware for opportunistic mobile social networks. *Pervasive and Mobile Computing, 11,* 148–167.

Baldauf, M., Dustdar, S., & Rosenberg, F. (2007). A survey on context-aware systems. *International Journal of Ad Hoc and Ubiquitous Computing, 2*(4), 263–277.

Bellotti, F., Berta, R., De Gloria, A., & Margarone, M. (2005). Implementing tour guides for travelers. *Human Factors and Ergonomics in Manufacturing & Service Industries, 15*(4), 461–476.

Biamino, G. (2011, March). Modeling social contexts for pervasive computing environments. In *Pervasive Computing and Communications Workshops (PERCOM Workshops), 2011 IEEE International Conference on* (pp. 415–420). IEEE.

Brézillon, P., & Gonzalez, A. J. (Eds.). (2014). Context in Computing: A Cross-Disciplinary Approach for Modeling the Real World. Springer.

Boongasame, L., Temdee, P., & Daneshgar, F. (2012). Forming buyer coalition scheme with connection of a coalition leader. *Journal of theoretical and applied electronic commerce research, 7*(1), 111–122.

Borriello, G. (2000). The challenges to invisible computing. *Computer, 11,* 123–125.

Bottazzi, D., Montanari, R., & Toninelli, A. (2007). Context-aware middleware for anytime, anywhere social networks. *IEEE Intelligent Systems, 22*(5), 23–32.

Budzik, J., & Hammond, K. J. (2000, January). User interactions with everyday applications as context for just-in-time information access. In *Proceedings of the 5th international conference on intelligent user interfaces* (pp. 44–51). ACM.

Chen, G., & Kotz, D. (2000). *A survey of context-aware mobile computing research* (Vol. 1, No. 2.1, pp. 2–1). (Technical Report TR2000-381). Dept. of Computer Science, Dartmouth College.

Chen, H., Finin, T., & Joshi, A. (2003). An ontology for context-aware pervasive computing environments. *The Knowledge Engineering Review, 18*(03), 197–207.

Cheverst, K., Davies, N., Mitchell, K., & Smith, P. (2000, January). Providing tailored (context-aware) information to city visitors. In *Adaptive Hypermedia and Adaptive Web-Based Systems* (pp. 73–85). Heidelberg: Springer Berlin.

Dey, A. K. (1998, March). Context-aware computing: the CyberDesk project. In *Proceedings of the AAAI 1998 Spring Symposium on Intelligent Environments* (pp. 51–54).

Dey, A. K., & Abowd, G. D. (2000a, June). The context toolkit: aiding the development of context-aware applications. In *Workshop on Software Engineering for wearable and pervasive computing* (pp. 431–441).

Dey, A. K., & Abowd, G. D. (2000b). Towards a better understanding of context and context-awareness. In *Workshop on the What, Who, Where, When, and How of Context Awareness, affiliated with the 2000 ACM Conference on Human Factors in Computer systems.*

Ejigu, D., Scuturici, M., & Brunie, L. (2007, April). Coca: a collaborative context-aware service platform for pervasive computing. In *Information Technology, 2007. ITNG'07. Fourth International Conference on* (pp. 297–302). IEEE.

Eubank, S., Guclu, H., Kumar, V. A., Marathe, M. V., Srinivasan, A., Toroczkai, Z., et al. (2004). Modelling disease outbreaks in realistic urban social networks. *Nature, 429*(6988), 180–184.

Fahy, P., & Clarke, S. (2004). CASS–a middleware for mobile context-aware applications. In Workshop on context awareness, MobiSys.

Fels, S., Sumi, Y., Etani, T., Simonet, N., Kobayshi, K., & Mase, K. (1998, March). Progress of C-MAP: a context-aware mobile assistant. In *Proceedings of AAAI 1998 Spring Symposium on Intelligent Environments* (pp. 60–67).

Finkelstein, A., & Savigni, A. (2001). A framework for requirements engineering for context-aware services.

Gu, T., Pung, H. K., & Zhang, D. Q. (2004, May). A middleware for building context-aware mobile services. In *Vehicular Technology Conference, 2004. VTC 2004-Spring. 2004 IEEE 59th* (Vol. 5, pp. 2656–2660). IEEE.

Gültekin, G., & Bayat, O. (2014). Smart location-based mobile shopping Android application. *Journal of Computer and Communications, 2*(08), 54.

Hofer, T., Schwinger, W., Pichler, M., Leonhartsberger, G., Altmann, J., & Retschitzegger, W. (2003, January). Context-awareness on mobile devices-the hydrogen approach. In *System Sciences, 2003. Proceedings of the 36th Annual Hawaii International Conference on* (pp. 10-pp). IEEE.

Hopper, A. (2000). The clifford paterson lecture, 1999. sentient computing. *Philosophical Transactions of the Royal Society of London A: Mathematical, Physical and Engineering Sciences*, 358(1773), 2349–2358.

Höpken, W., Fuchs, M., Zanker, M., & Beer, T. (2010). Context-based adaptation of mobile applications in tourism. *Information Technology & Tourism, 12*(2), 175–195.

Intayoad, W., Becker, T., & Temdee, P. (2017). Social Context-Aware Recommendation for Personalized Online Learning. *Wireless Personal Communications*, 1–17.

Kabir, M. A., Han, J., Yu, J., & Colman, A. (2012, June). SCIMS: a social context information management system for socially-aware applications. In *International Conference on Advanced Information Systems Engineering* (pp. 301–317). Heidelberg: Springer Berlin.

Kabir, M. A., Colman, A., & Han, J. (2014a). SocioPlatform: a platform for social context-aware applications. In *Context in Computing* (pp. 291–308). New York: Springer.

Kabir, M. A., Han, J., & Colman, A. (2014b). Sociotelematics: Harnessing social interaction-relationships in developing automotive applications. *Pervasive and Mobile Computing, 14,* 129–146.

Kabir, M. A., Han, J., Yu, J., & Colman, A. (2014c). User-centric social context information management: an ontology-based approach and platform. *Personal and Ubiquitous Computing, 18*(5), 1061–1083.

Khor, T. L. (2016). Personal Shopping Assistant. *Doctoral dissertation*. UTAR.

Korpipää, P., Jani, M., Kela, J., & Malm, E. J. (2003). Managing context information in mobile devices. *IEEE Pervasive Computing, 3,* 42–51.

Kourtellis, N., Finnis, J., Anderson, P., Blackburn, J., Borcea, C., & Iamnitchi, A. (2010, November). Prometheus: User-controlled p 2p social data management for socially-aware applications. In *Proceedings of the ACM/IFIP/USENIX 11th International Conference on Middleware* (pp. 212–231). Springer-Verlag.

Li, J., & Dabek, F. (2006, February). F2F: Reliable storage in open networks. In *IPTPS*.

Liang, G., & Cao, J. (2015). Social context-aware middleware: a survey. *Pervasive and Mobile Computing, 17,* 207–219.

Loke, S. (2006). Context-aware pervasive systems: architectures for a new breed of applications. CRC Press.

Lu, Y., Tsaparas, P., Ntoulas, A., & Polanyi, L. (2010, April). Exploiting social context for review quality prediction. In *Proceedings of the 19th international conference on World wide web* (pp. 691–700). ACM.

Manyika, J., Chui, M., Brown, B., Bughin, J., Dobbs, R., Roxburgh, C., et al. (2011). Big data: the next frontier for innovation, competition, and productivity.

Mislove, A., Gummadi, K. P., & Druschel, P. (2006, August). Exploiting social networks for internet search. In *5th Workshop on Hot Topics in Networks (HotNets06)*. Citeseer (p. 79).

Modes, B. (2012). From context awareness to socially aware computing.

Norman, D. (1998). The invisible computing.

Pascoe, J., Ryan, N., & Morse, D. (1999, January). Issues in developing context-aware computing. In *Handheld and ubiquitous computing* (pp. 208–221). Heidelberg: Springer Berlin.

Perera, C., Zaslavsky, A., Christen, P., & Georgakopoulos, D. (2014). Context aware computing for the internet of things: a survey. *IEEE Communications Surveys & Tutorials, 16*(1), 414–454.

Schilit, B. N., Adams, N., Gold, R., Tso, M. M., & Want, R. (1993, October). The PARCTAB mobile computing system. In *Workstation Operating Systems, 1993. Proceedings., Fourth Workshop on* (pp. 34–39). IEEE.

Schilit, B., Adams, N., & Want, R. (1994, December). Context-aware computing applications. In *Mobile Computing Systems and Applications, 1994. WMCSA 1994. First Workshop on* (pp. 85–90). IEEE.

Schuster, D., Rosi, A., Mamei, M., Springer, T., Endler, M., & Zambonelli, F. (2013). Pervasive social context: taxonomy and survey. *ACM Transactions on Intelligent Systems and Technology (TIST), 4*(3), 46.

Sheng, Q. Z., Benatallah, B., Maamar, Z., Dumas, M., & Ngu, A. H. (2004, January). Enabling personalized composition and adaptive provisioning of web services. In *Advanced Information Systems Engineering* (pp. 322–337). Heidelberg: Springer Berlin.

Strang, T., & Linnhoff-Popien, C. (2004, September). A context modeling survey. In *Workshop Proceedings*.

Tennenhouse, D. (2000). Proactive computing. *Communications of the ACM, 43*(5), 43–50.

Toninelli, A., Pathak, A., & Issarny, V. (2011, May). Yarta: A middleware for managing mobile social ecosystems. In *International Conference on Grid and Pervasive Computing* (pp. 209–220). Heidelberg: Springer Berlin.

Wang, M. M., Cao, J. N., Li, J., & Dasi, S. K. (2008). Middleware for wireless sensor networks: A survey. *Journal of computer science and technology, 23*(3), 305–326.

Want, R., Hopper, A., Falcao, V., & Gibbons, J. (1992). The active badge location system. *ACM Transactions on Information Systems (TOIS), 10*(1), 91–102.

Weiser, M. (1991). The computer for the 21st century. *Scientific American, 265*(3), 94–104.

Wu, H., & Natchetoi, Y. (2007, May). Mobile shopping assistant: integration of mobile applications and web services. In *Proceedings of the 16th international conference on World Wide Web* (pp. 1259–1260). ACM.

Yu, Z., Yu, Z., & Zhou, X. (2012). Socially aware computing. *Chinese Journal of Computers, 35*(1), 16–26.

Chapter 2
Context and Its Awareness

Abstract This chapter aims to introduce the definition of context and its awareness, which are shown through the review of existing works from the early and the current state of context-aware computing. The categories, the characteristics, and the property of context are described. The context-aware architecture is discussed in this chapter. The common components of context-aware applications can be summarized into three components including perceiving component, thinking component, and acting component. Additionally, some perspectives for context definition as social context, context categorization and context awareness as personalized and rationalized awareness are emphasized at the end of this chapter.

Understanding context is crucial because it has the main role in executing the context-aware applications. Basically, it is necessary to understand the definition of context, its categories, its characteristics, its properties, and its awareness in order to obtain proper design and development. This chapter will be discussing all these issues including some perspectives relating to the context for the future requirements and applications.

2.1 Context Definition

For the last decades, there is a significant amount of prototypes, systems, and applications implementing context-aware computing concept with variety kinds of contexts. Before going into detail of how to develop context-aware applications, it is important to understand the definition of context and its evolution. According to the Cambridge Dictionary Online,[1] the context is defined as "the situation within which something exists or happens, and that can help explain it." The word "within" simply shows something inherently influences something to happen. Therefore, the definition of context-aware computing relates directly to the definition of something having the assertion as a person, a circumstance or a computer

[1]http://dictionary.cambridge.org/.

P. Temdee and R. Prasad, *Context-Aware Communication and Computing: Applications for Smart Environment*, Springer Series in Wireless Technology, DOI 10.1007/978-3-319-59035-6_2

system. The context-aware applications are the systems that can adapt their operations or behaviors to the current contexts with or without the explicit intention of the user intervention. The context is thus important as it has the primary role in executing the application. Most of the early research works in context-aware computing aimed to identify what contexts are used in their applications. As mentioned before, the history of context-aware applications started when the Active Badge Location System was introduced (Want et al. 1992). This system could determine the current location of the users and forwarded the calls to the phone closest to the users. The user's location was detected by infrared technology. Consequently, the location was only the context executing the response from the system. At that time, the location was frequently used particularly for any location-aware application especially tour guide applications (Abowed et al. 1997; Sumi et al. 1998; Cheverst et al. 2000).

Once there was a diversity of context-aware applications, more entities were introduced as the contexts. For example, Schilit and Theimer (1994) described context as locations, identities of nearby people, objects, and changes to those objects. Ryan et al. (1999) defined the context as the user's location, environment, identity, and time. Dey (1998) described the context as the user's emotional state, location, orientation, date, and time, as well as objects and people in the environment. Some works use synonyms for context, such as environment and situation (Brown 1995; Franklin and Flaschbart 1998; Rodden et al. 1998; Hull et al. 1997; Ward et al. 1997; Abowd and Mynatt 2000). It can be seen that those works provide the definitions of context that are apparently based on the examples and synonyms. Identifying the general description for context is entirely challenging.

As claimed by Dey (1998), the context definitions at the early stage were too specific, and it could not be used to identify the other contexts in a broader sense. More general definition of context had been introduced consequently (Brown 1995; Pascoe 1998; Dey 2001; Dourish 2004; Bazire and Brézillon 2005; Zimmermann et al. 2007; Jumisko-Pyykkö and Vainio 2012; Alshaikh and Boughton 2013; Perera et al. 2014). For example, Brown (1995) defined the context as the elements of the user's environment which the computer knows about it. Pascoe (1998) introduced the context as a subjective concept that is defined by the entity that perceives it. It could be described as the subset of physical and conceptual states of interest to a particular entity. Dey and Abowd (2000) and Dey (2001) defined the context as "any information that can be used to characterize the situation of an entity" where "an entity is a person, place, or object that is considered relevant to the interaction between a user and an application, including the user and applications themselves." Closely related to the notion of context is the concept of the situation. The relationship between context and situation is illustrated by Dey (1998) that the situation of entity can be determined by the aggregation of context information. In this sense, the situation can be viewed as the higher abstract of context. In conclusion, many works seem to agree with the definition of context for some aspects, although the consensus of general definition of context is still diverse (Baldauf et al. 2007; Alegre et al. 2016). Some common agreements are, for example, the context can be defined as any information for characterizing the situation of any entity. The context can give meaning to something else by itself or combination with other contexts. Moreover, the context typically means the operational term rather than its inherent properties.

2.2 Context Categories and Characteristics

Not only is the conceptual definition of context diverse, but also the categories and characteristics. Category and characteristics of context are necessary for the application or service designers to discover the context of their applications and services. This section shows the variety of context categories and characteristics used for some various applications. The context characterized as location, identity, time, and activity (Schilit and Theimer 1994) is normally used for describing the situation of a particular entity. The entity can be the place, people or things. These context types not only simply answer the questions of who is doing what, when, and where, but also leads to other sources of contextual information. For example, the personal identification number can provide the other related information such as affiliation, addresses, date of birth, etc. The entity's location can determine other objects or people nearby and what activity is occurring nearby. Therefore, the context can be considered as primary context and secondary context (Perera et al. 2014) where the primary context can be used to find the secondary context of the same entity. More specifically, primary context is any information obtained from the sensors and retrieved without using existing context and and any data fusion. On the other hands, secondary context is any information that is derived by the manipulation of primary context. For example, the distance between two sensors with particular data fusion is called secondary context while the sensor data from each sensor is called primary context. Moreover, the retrieved context such as friend list, email address, etc. are also known as secondary context while the name of the user is considered as primary context. For another example, Dey et al. (2001) proposed that the context can be categorized into four categories including identity, location, status, and time. Identity means that each entity has a unique identifier. Location means the entity's position. Status means the intrinsic properties of the entity such as the temperature in the room, the lightness in the car, etc. The status is also considered as the activity. Finally, time is used to define the situation accurately. These ways to characterize the context cannot cover all emerged context types due to the variety of system requirements.

There are many ways to classify context into different categories. For example, the work proposed by Chen and Kotz (2000) defined the categories of the context as computing context, user context, physical context, temporal context, and context history. The computing context includes network connectivity, communication bandwidth, and local computing resources such as printers, displays, etc. User context can be user profile, location, social situation. Physical context can be lighting and noise levels, traffic conditions, and temperature. Temporal context includes time of day, week, month, and season of the year. Context history is the storage of existing context in different points of time. Another popular way to classify types of context is to classify it into external and internal context (Baldauf et al. 2007; Schuster et al. 2002; Prekop and Burnett 2003). At the same time, they can be called physical and logical contexts respectively (Hofer et al. 2003). The physical or external context refers to the context that can be measured by hardware sensors such as location, light, sound, temperature, etc. On the other hand, the logical or internal context is something specified by the user such as user's goal, task, etc. Henricksen (2003) proposed that the context can be categorized as sensed, static, profiles, and derived context categories. Sensed context is sensor data directly detected by the sensors such as temperature, humidity, speed, etc. Static context is the information that does not change over time such as the identification of sensors from the manufacturer, person identification, etc. Profile context means the information that can evolve over time with low frequency such as the location of the sensor, the status of the person, etc. Finally, the derived context means the information that is computed by using primary context such as the distance between two sensors. Other popular types of context are the operational and the conceptual context (Van Bunningen et al. 2005; Alegre et al. 2016). The operational context means context relating to system's operation and it involves directly to the context acquisition, modeling, and treating. At the same time, the conceptual context covers meaning and relationship and can explain the relationship between contexts.

Nowadays, social context has become popular because there have been the increasing demands of social-aware applications. Social context is used to be defined as the person nearby or the group to which the user belongs. Recently, new definitions of social context are proposed. For example, Liang and Cao (2015) defined the social context as "a set of information derived from direct or indirect interactions among people in both virtual and physical world." Social context plays a significant role in public security and public health as automatic crowd detection, criminal analysis, disease infection, etc. (Eubank et al. 2004). Most of the social-aware applications deal with a large of digital traces of the users. Moreover, the application itself has shifted into networked system interacting with the community rather than single user perspective system (Eubank et al. 2004).

Besides characterizing context from what context is used for the applications, the context can also be classified from acquisition ways. For this point of view, the context can be divided into 2 different types including state information and change event. For state information, the application actively requests (pull) required context and accesses to actual and historical data such as current location, device, etc. For the change event, the application registers for particular change events and waits passively for the events. Then, the context service notifies registered applications about changes of state (push) such as the location changes, network changes, etc. For example, the air conditioner is set to turn on if the temperature is more than 25 ° C means that the system uses the stat status to turn on the air conditioner. At the same time, the air conditioner is set to turn on if the temperature increases means that the system uses the change event status to turn on the air conditioner.

2.3 Context Property

Context property is essential for system design and development for context-aware applications. Different properties require different detail for designing and development. Context attributes can have the variety of properties. For example, time-dependent context which represents dynamic information that the values change over time. At the same time, static information like date-of-birth can be interpreted as information with change frequency of zero. Historic context is the context representing values at different points in time. Incorrect context is the context that can be incorrect due to inaccurate sensor information, measurement failure, wrong assumptions for derivation and interpretation. The quality of context depends on uncertainty in measurements and many evitable reasons. Multiple-resource context means that the same information can be gathered in different ways such as the location of a person can be collected from GPS, the position of the device, etc. Multidimensional/Heterogeneous context means that the context can be physical or technical context and private or social context at the same time. Distributed context means that the context occurs everywhere and all the time. Insecure context means that some contexts require security and privacy protections. Imperfectcontext means that the context always is imperfect especially regarding incomplete and inconsistent. Unforeseeable context means that the unforeseen context always occurs in the real life system.

2.4 Context Awareness

Context awareness represents the ability of the system that can use the context to provide the appropriate response to the users. Many systems can be considered as the context awareness systems, but they are called by other names such as smart system, intelligent system, adaptive system, etc. Since there is the variety of systems that can be considered as the context-aware system and the diversity of the context definition, it is also still challenging to make the consensus of the definition of context awareness nowadays. The term context awareness was firstly called sentient (Schilit and Theimer 1994) and later defined by Dey (2001) as "A System is context-aware if it uses context to provide relevant information and/or services to the user, where relevancy depends on the user's task." Since the term context awareness highly depends on the definition of context, the context awareness can be diverse. Instead of paying much effort into clarifying definitions, many works illustrates the meaning by implementing the system with different levels of awareness.

The context awareness can be identified into 3 levels based on the user interaction (Barkhuus and Dey 2003) including personalization, passive context-awareness, and active context-awareness. Personalization means that the system allows the users to set their preferences to the system manually. For example, the users can set their preferred coffee taste where the coffee machine can maintain the prefered taste for the users. Passive context-awareness means that the system continuously monitors the environment and proposes the appropriate choices to the users for taking the actions. For example, the promotion or discount messages are sent to the user's browser when they are surfing the Internet and wait for the action from the users. Finally, active context-awareness means that the system continuously and autonomously monitors the situation and acts independently without the intervention from the user. For example, the smart home can watch the intruders and will autonomously notify the owners or the police when someone, not the owner, has broken into the house.

Once the context-aware applications have to deal with more complexity of context, the context-aware application itself requires a deeper understanding of context which involves the complex relation of entities (Alegre et al. 2016). Therefore, the context-aware system can be classified into 2 different modes based on the interaction with the system including execution and configuration modes. Execution mode refers to the system acts or behaves particularly for a particular situation. For example, when the phone receives the call during the meeting, the phone turns automatically to the silent mode. At the same time, configuration mode refers to the adjustment of action or behavior that the system will be performing in the future. For example, the phone can adjust the action accordingly to the user's preference that the phone will not turn to silent mode when receiving the calls from someone. At the same time, the context-aware system can be considered as the active or passive system. The active system means that the system changes its content automatically while the passive system means that the system will change the content when the user has explicit involvement. Alegre et al. (2016) also suggested that context-aware system does not have to be completely active or passive. They can be some degrees in between as called hybrid mode. Following Alegre et al. (2016), there are possible 4 types of context-aware systems accordingly to the interaction with and without the involvement of the users.

Active execution

The system acts automatically depending on the context and can be called as a self-adaptive system. It can adjust its behavior accordingly to the perception of the environment and itself. For example, the screen of the smart device can switch between landscape and portrait depending on its pose automatically. The air conditioner turns on automatically when the room temperature is higher than the particular degree. This system requires less or no effort from the user and no special knowledge to use the system. However, it's hard to ensure that the system will perform appropriate actions or behaviors. At the same time, the users can be uncomfortable because they do not know what their information will be used. The examples of existing systems are context-aware self-adaptive frameworks for mobile application (Cheng et al. 2009; Salehie and Tahvildari 2009; Mizouni et al. 2014), MUSIC project (Hallsteinsen et al. 2012; Rouvoy et al. 2009; Geihs and Wagner 2012), etc.

Passive execution

This system requires the involvement of the users that they have to specify how the application should change or behave in some particular situations. The system can provide all possible information or actions to the user to select the appropriate actions. This system gains the trust from the users because they understand how the system works. The system is designed to take the actions that the user wants, so it is easy to evaluate the system's behavior. However, this system requires higher context understanding and the development of explanation generation which are difficult to develop. At the same time, the system needs more information to explain actions. The examples of existing systems belonging to this mode are some tool-kits for context-aware applications (Lim and Dey 2009).

Active configuration

This system can learn from the user preferences for automatically change or evolve the rules for the future behavior. This system requires less or no effort from the user. At the same time, the user requires no special knowledge to use the system. The system is expected to understand the behavior or the habits of the users. It's hard to determine which rules should be added or deleted for particular changes. The system needs the complex module to deal with inaccurate and uncertain sensed data. The existing systems are, for instance, the context-aware adaptive systems having foreseen and unforeseen types of context (Mori 2011; Inverardi and Mori 2013), the system discovering patterns within the user actions (Aztiria et al. 2013), the system that can generate reasoning rules automatically appropriately (Ibarra et al. 2014), etc.

Passive configuration

This system requires the involvement of the users by manually providing the personalized information to the system such as preferences, likes, expectation, etc. This system offers not only greater control and ownership to the users but also the greater creativity and motivation. It can release the burden of the developer because the users can manage their task, and they know their task the best. However, the users might be forced to contribute and cooperate with something they could lack experience. At the same time, the system may have to deal with more sophisticated methods. The examples of existing systems are Trigger-action programming (Ur et al. 2014; Huang and Cakmak 2015), iCap project (Dey et al. 2006) which is a system that is the intermediate layer between low-level toolkits and users, etc.

For our perspective on context awareness, it is a combination of some degree of the system acting personally and rationally with and without the intervention from the users. System acting personally means that the system would like to act in the manner of user's preference with or without the intervention from the users. On the other hand, system acting rationally means that the system can adapt itself to be able to interact with the new environment appropriately with or without the intervention from the users. It can be seen that, both systems can be triggered with or without the interventions from the users. The detail of this perspective will be later discussed in the last section of this chapter.

2.5 Context-Aware Architecture

Context-aware systems can be implemented in many different ways under many considerations such as individual requirements, the number of users, types of user devices, context acquisition methods, etc. The system architecture is necessary required for system representation and implementation. The system architecture is an abstraction which is used for generalizing the systems without showing the detail of implementation (Alegre et al. 2016). Like any other systems, the context-aware system requires flexible system architecture. The existing works illustrate the evolution of context-aware architectures supporting from specific purpose application to the general purpose application.

Winograd (2001) introduced three different context-aware architecture including widgets, networked services, and blackboard model. The widget is a software component providing the interface for hardware sensors (Dey and Abowd 2001). By hiding the low-level detail of sensing, it is easy to develop the application and obtain reusability. The widget can increase the efficiency but may not be robust for general purpose architecture. Many widgets are controlled by the widget manager. At the same time, the networked services can be considered as more flexible approach. Instead of having widget manager, the networked services can be found by using particular discovery techniques. Finally, the blackboard model represents the data-centric view. The blackboard is a shared media for notifying when some specified event happen. The idea came from many experts sitting around a blackboard and cooperating to solve a particular problem together (Taylor et al. 2009). This architecture is easy to implement, but there is the need of the centralized server to host the blackboard.

There have been other different points of view for proposing system architecture for context-aware applications. For example, three different architectures can be classified based on context acquisition approaches (Chen 2004) including direct sensor access, middleware infrastructure, and context server. For direct sensor access, this approach is used for the devices that have built-in sensors. The software gathers information directly from the sensors. This method can be considered as a tightly coupled approach that may not be suitable for distributed systems. For middleware architecture, it is frequently used by modern software design using encapsulation to separate functionality. The middleware approach introduces a layered architecture for hiding sensing detail at low-level. This approach promotes reusability of hardware sensors and extensibility of the system. For context server, this approach introduces remote management component to the middleware-based architecture. The server is mainly responsible for gathering sensor data by facilitating concurrent multiple accessing. The advantage of this approach is to promote the reusability of sensors and the decrement of resource intensive operation.

.Three different architectures can also be classified based on actions required by the context-aware applications (Hu et al. 2008) including acquisition, representation, delivery and reaction of the contexts. The architectures thus include no application-level context model, implicit context model, and explicit context model respectively. No application-level context model means that the applications perform all actions within the application boundaries. Implicit context model means that the applications use some other resources to carry out all actions such as libraries, frameworks, toolkits, etc. Explicit context model means that the applications use a context management infrastructure or middleware solution for performing all actions outside the application boundary. Context management and application are separated for being developed and extended independently.

The layered architectures have been proposed to satisfy the needs of general architecture that can be tailored to any application appropriately. The particular characteristic of layered architectures is that the functionalities are divided into layers and the components in a meaningful manner. Each component performs a limited task independently to support the extensibility. The well-known layered architecture for the context-aware system was proposed by Baldauf et al. (2007) having five different layers including sensors, raw data retrieval, storage and management, pre-processing, and application layer.

Sensor layer deals with a collection of various sensors including physical, virtual and logical sensors (Indulska and Sutton 2003). Physical sensors for almost every physical measurement are widely available nowadays. For example, location can be sensed by using GPS, Global System for Mobile Communication (GSM) or satellite system. Light can be detected by photodiodes. Temperature can be detected by thermometers, etc. Next, virtual sensors mean the source of context data from software applications or services. For example, the location can be identified by browsing travel booking system besides using only physical sensors as location tracking system. Finally, logical sensors combine some physical and virtual sensors together with additional information to obtain higher level abstraction. For example, the location can be detected by analyzing user login and the mapping of their device locations. At the same time, the logical sensors can be considered as the fusion of physical and virtual sensors. As shown with the name, raw data retrieval layer deals with retrieval of raw context data by using appropriate drivers for physical sensors and Application Programming Interface (API) for virtual and logical sensors. Pre-processing layer is used for preparing the information ready for the applications. Storage and management layer deal with organizing the gathered data, keeping them in the appropriate space and offering them to the clients. There are two ways of data accessing from the clients including synchronous and asynchronous modes. For synchronous mode, the client sends a message to request the data until receiving the answers from the server. For asynchronous mode, the client subscribes to specific interest events which the client will be simply notified when the event happens. Finally, the application layer is responsible for implementation of the applications.

The system architecture can also be implicitly represented by considering the integrated features in any context-aware application. For example, Abowd et al. (1999) identified three features that the context-aware application could support including presentation, execution, and tagging. Presentation means that the context can decide what information or the services should be presented to the users. The general example is the advertisement message sent to the users when they are in the department store through their smartphones (Institutes 2011). The smart refrigerator (Moses 2012) connects to the mobile and informs what the users should bring into home from the department store. These examples illustrate the idea of selecting suitable methods for representing the appropriate information to the users. Next, the execution means that the actions are taken automatically based on the context. For example, the air-conditioner is turned on when the users start to drive home from somewhere else. Tagging means that the collected sensor data needs to be analyzed, fused, and interpreted so that it can be processed and understood later. Context tagging can also be called context annotation.

Although many works propose different architectures for the context-aware application, those designs frequently share the common ideas. Some similar ideas are, for example, promoting many extraordinary abilities such as the heterogeneity of sensor sources, scalability of the sensors, tractability for controlling and debugging, providing tolerance for the component failure, supporting privacy, promoting mobility of users and applications, enabling ease of development and configuration, etc.

2.6 Common Components

So far, it can be clearly seen that the context-aware applications can respond to any stimuli like other living things or artifacts (Loke 2006) such as the robot, software agents, etc. Especially for context-aware systems, identifying, understanding, and exploiting the context of entities are the primary tasks. Accordingly to Loke (2006), this section describes the standard components among context-aware applications including perceiving or sensing component, thinking component, and acting component.

2.6.1 Perceiving Component

Perceiving or sensing is the acquisition of data or information about the physical world which is used by a computer system to determine appropriate actions. It can be both biological and non-biological sensors. Multiple sensors may give a more comprehensive view of the physical world but perhaps more complexity of data manipulation. Information can be sensed through many different sensors such as light sensor, temperature sensors, motion sensors, touch sensors, etc. There are some concerns about sensors. For example, the sensors could be embedded in the environment, and sometimes they can be worn unobtrusively. It is challenging to determine the best reasoning and combination methods for acquiring context information for particular context-aware applications.

2.6.2 *Thinking Component*

The critical component of the context-aware application is to make use of gathered information from the sensors to perform the appropriate response to the users and/or the environment. The thinking component aims to make sense from all collected information. There are two schools of thought widely accepted including rationalists and empiricists. Rationalists use only reasoning to gain knowledge. On the other hand, empiricists used experience through the senses and stored in the memory to acquire knowledge. The combination of two methods is also accepted in which some information is perceived via the sensors and employs reasoning to infer more knowledge. Context-aware applications acquire sensor information and then reason it with other knowledge so that the further knowledge can be assumed. There have been many ways of making sense from information, for example, using mathematical models, using feature-based inference techniques such as pattern recognition, and neural networks, and using cognitive-based models as knowledge bases, and fuzzy logic.

2.6.3 *Acting Component*

Once context information has been gathered, the context is analyzed, and the situations are recognized, the appropriate actions are expected to be taken. Not only is the performance the primary consideration, but also the control action such as override actions, cancel actions or stop actions. The context-aware applications need the appropriate design of the responses of those control actions. Moreover, there have also been varieties of actuators that the application needs to interact with their environment and users. The most concern for acting component is to select the appropriate actuators to satisfy the requirements of the users and the enviornment which are typically domain-specific applications.

2.7 Common Architecture

It can be seen that all proposed context aware architectures attempt to represent the architecture with different perspectives. However, three primary components can be identified including perceiving, thinking, and acting components respectively. For example, Fig. 2.1 shows that the layered architecture proposed by the works of Baldauf et al. (2007) and Abowd et al. (1999) share three common components.

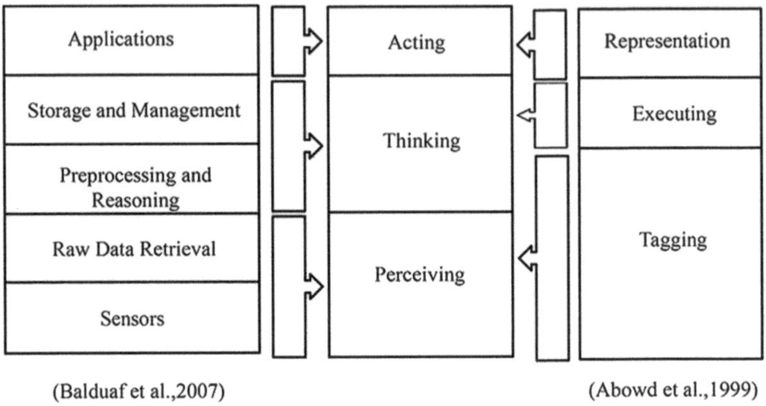

Fig. 2.1 Common components of context-aware architectures

2.8 Perspectives of Context

This section shows the perspectives for context definition, context categorization, and context awareness of this book.

2.8.1 Definition Perspective

As it can be seen from the last section, it is quite clear from the literature that it is not easy to define the definition of context because of variety kinds of context-aware applications. At the early stage, the contexts are easily defined accordingly to what contexts are used in particular applications such as location, time, environment, identity, etc. All among those contexts, location is the frequently used at the early stage. Then more contexts are introduced later such as emotional state, orientation, social context, etc. Later, there has been the effort to define the conceptual definition for context. Most of the research works do agree that the conceptual definition of context is still complicated to be built even nowadays. Figure 2.2 shows the evolution of context definition that the conceptual definition is still going on and new contexts are on the way to be announced.

2.8.2 Categorization Perspective

As the different perspective on context, it can be concluded that there is no single categorization can accommodate all types of context nowadays. For our perspective on the category of the context, the context can be classified as the individual and

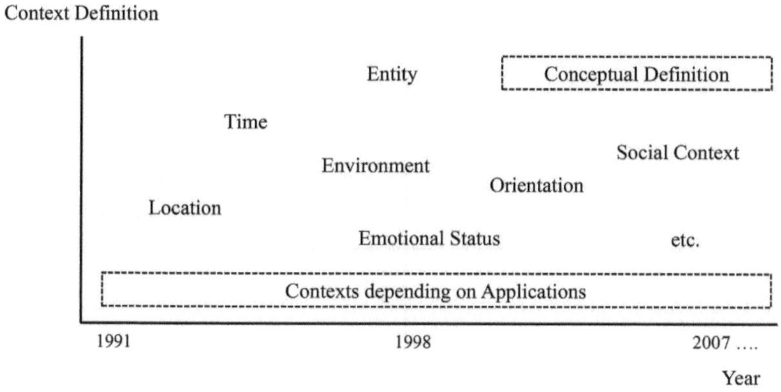

Fig. 2.2 Evolution of context definition

the social context. The individual context is the context that provides the meaning without the interaction with other entities while the social context is the context that needs interaction among entities to provide the meaningful information. The interactions can be the peer or group interactions. Absolutely, the entity can be anything not only people (Liang and Cao 2015) such as sensors, software agents, insects, robots, even the machines as on Internet of Thing (IoT) paradigm. Moreover, each type of context can have its attributes which can be tailored for any applications individually. Figure 2.3 shows the conceptual diagram of individual and social context based on the proposed perspective.

From Fig. 2.3, the interaction between entities does not simply mean the fusion or the aggregation between or among the contexts. For example, to determine the group information which is the social context, it is not just aggregating individual contexts (I1,I2,I3) together. On the other hand, the social context (C1,C2,C3) can be identified by the manipulations of all interactions among all entities (E1,E2,E3) instead. The future context-aware applications requiring the social context with this perspective will be discussed again in Chap. 6.

Fig. 2.3 Context categorization based on context interaction

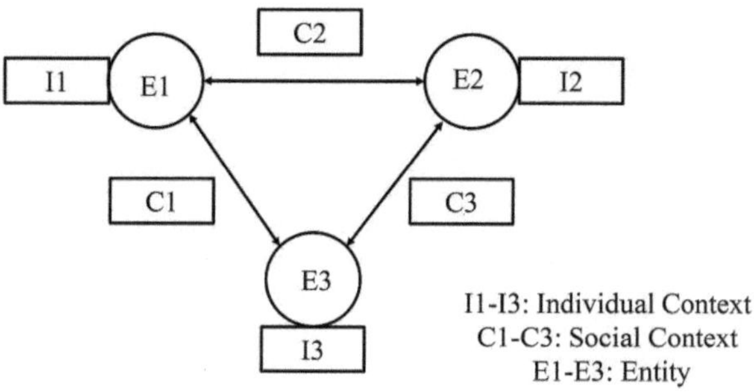

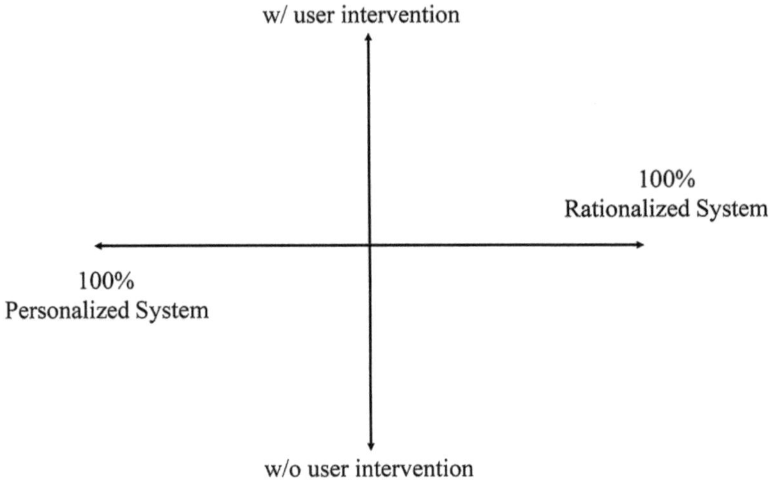

Fig. 2.4 Personalized and rationalized context-aware application

2.8.3 Awareness Perspective

As mentioned before, the context awareness can be considered as the systems acting personally or rationally with and without the intervention from the users. This book would like to identify the awareness model as in between the personal based reflex model and rational based reflex model as shown in Fig. 2.4.

It can be seen from Fig. 2.4 that any system can have the combination of personalized and rationalized systems. Both systems can be executed with or without the intervention from the users. The personalized system means that the system can act as the user's manner no matter what the users have trained the system consciously. On the other hand, the rationalized system means that the system can act appropriately to the situation or environment no matter what the users have trained the system consciously.

References

Abowd, G. D., Atkeson, C. G., Hong, J., Long, S., Kooper, R., & Pinkerton, M. (1997). Cyberguide: A mobile context-aware tour guide. *Wireless Networks, 3*(5), 421–433.

Abowd, G. D., Dey, A. K., Brown, P. J., Davies, N., Smith, M., & Steggles, P. (1999, January). Towards a better understanding of context and context-awareness. In *Handheld and ubiquitous computing* (pp. 304–307). Berlin, Heidelberg: Springer.

Abowd, G. D., & Mynatt, E. D. (2000). Charting past, present, and future research in ubiquitous computing. *ACM Transactions on Computer—Human Interaction (TOCHI), 7*(1), 29–58.

Alegre, U., Augusto, J. C., & Clark, T. (2016). Engineering context-aware systems and applications: A survey. *Journal of Systems and Software, 117,* 55–83.

Alshaikh, Z., & Boughton, C. (2013, October). Notes on synthesis of context between engineering and social science. In *International and Interdisciplinary Conference on Modeling and Using Context* (pp. 157–170). Berlin, Heidelberg: Springer.

Aztiria, A., Augusto, J. C., Basagoiti, R., Izaguirre, A., & Cook, D. J. (2013). Learning frequent behaviors of the users in intelligent environments. *IEEE Transactions on Systems, Man, and Cybernetics: Systems, 43*(6), 1265–1278.

Baldauf, M., Dustdar, S., & Rosenberg, F. (2007). A survey on context-aware systems. *International Journal of Ad Hoc and Ubiquitous Computing, 2*(4), 263–277.

Barkhuus, L., & Dey, A. (2003, October). Is context-aware computing taking control away from the user? Three levels of interactivity examined. In *International Conference on Ubiquitous Computing* (pp. 149–156). Berlin, Heidelberg: Springer.

Bazire, M., & Brézillon, P. (2005, July). Understanding context before using it. In *International and Interdisciplinary Conference on Modeling and Using Context* (pp. 29–40). Berlin, Heidelberg: Springer.

Brown, P. J. (1995). The stick-e document: A framework for creating context-aware applications. *Electronic Publishing-Chichester-, 8,* 259–272.

Chen, G., & Kotz, D. (2000). A survey of context-aware mobile computing research (Vol. 1, No. 2.1, pp. 2-1). Technical Report TR2000-381, Department of Computer Science, Dartmouth College.

Chen, H. (2004). *An intelligent broker architecture for pervasive context-aware systems.* Baltimore County: University of Maryland.

Cheng, B. H., de Lemos, R., Garlan, D., Giese, H., Litoiu, M., Magee, J., & Taylor, R. (2009, May). Seams 2009: Software engineering for adaptive and self-managing systems. In *Proceedings of the 2009 31st International Conference on Software Engineering: Companion Volume* (pp. 463–464). IEEE Computer Society.

Cheverst, K., Davies, N., Mitchell, K., Friday, A., & Efstratiou, C. (2000, April). Developing a context-aware electronic tourist guide: Some issues and experiences. In *Proceedings of the ACM, SIGCHI conference on Human Factors in Computing Systems* (pp. 17–24).

Dey, A.K. (1998, March). Context-aware computing: The CyberDesk project. In *Proceedings of the AAAI 1998 Spring Symposium on Intelligent Environments* (pp. 51–54).

Dey, A. K., & Abowd, G. D. (2000). Towards a better understanding of context and context-awareness. In *Workshop on the What, Who, Where, When, and How of Context Awareness, affiliated with the 2000 ACM Conference on Human Factors in Computer systems.*

Dey, A. K. (2001). Understanding and using context. *Personal and Ubiquitous Computing, 5*(1), 4–7.

Dey, A. K., Abowd, G. D., & Salber, D. (2001). A conceptual framework and a toolkit for supporting the rapid prototyping of context-aware applications. *Human-Computer Interaction, 16*(2), 97–166.

Dey, A.K., Sohn, T., Streng, S., & Kodama, J. (2006, May). iCAP: Interactive prototyping of context-aware applications. In *International Conference on Pervasive Computing* (pp. 254–271). Berlin, Heidelberg: Springer.

Dourish, P. (2004). What we talk about when we talk about context. *Personal and Ubiquitous Computing, 8*(1), 19–30.

Eubank, S., Guclu, H., Kumar, V. A., Marathe, M. V., Srinivasan, A., Toroczkai, Z., et al. (2004). Modelling disease outbreaks in realistic urban social networks. *Nature, 429*(6988), 180–184.

Franklin, D., & Flaschbart, J. (1998, March). All gadget and no representation makes jack a dull environment. In *Proceedings of the AAAI 1998 Spring Symposium on Intelligent Environments* (pp. 155–160).

Geihs, K., & Wagner, M. (2012, November). Context-awareness for self-adaptive applications in ubiquitous computing environments. In *International Conference on Context-Aware Systems and Applications* (pp. 108–120). Berlin, Heidelberg: Springer.

Hallsteinsen, S., Geihs, K., Paspallis, N., Eliassen, F., Horn, G., Lorenzo, J., et al. (2012). A development framework and methodology for self-adapting applications in ubiquitous computing environments. *Journal of Systems and Software, 85*(12), 2840–2859.

Henricksen, K. (2003). *A framework for context-aware pervasive computing applications*. Queensland: University of Queensland.
Hofer, T., Schwinger, W., Pichler, M., Leonhartsberger, G., Altmann, J., & Retschitzegger, W. (2003, January). Context-awareness on mobile devices-the hydrogen approach. In *System Sciences, 2003. Proceedings of the 36th Annual Hawaii International Conference on IEEE* (p. 10-pp).
Hu, P., Indulska, J., & Robinson, R. (2008, March). An autonomic context management system for pervasive computing. In *Pervasive Computing and Communications, 2008. PerCom 2008. Sixth Annual IEEE International Conference on IEEE* (pp. 213–223).
Huang, J., & Cakmak, M. (2015, September). Supporting mental model accuracy in trigger-action programming. In *Proceedings of the 2015 ACM International Joint Conference on Pervasive and Ubiquitous Computing* (pp. 215–225). ACM.
Hull, R., Neaves, P., & Bedford-Roberts, J. (1997, October). Towards situated computing. In *Wearable Computers, 1997. Digest of Papers., First International Symposium on IEEE* (pp. 146–153).
Ibarra, U.A., Augusto, J.C., & Goenaga, A.A. (2014, June). Temporal reasoning for intuitive specification of context-awareness. In *Intelligent Environments (IE), 2014 International Conference on IEEE* (pp. 234–241).
Indulska, J., & Sutton, P. (2003, January). Location management in pervasive systems. In *Proceedings of the Australasian information security workshop conference on ACSW frontiers 2003-Volume* 21 (pp. 143–151). Australian Computer Society, Inc.
Institutes, C. (2011). Smart networked objects and internet of things. *Carnot Institutes' Information Communication Technologies and Micro Nano Technologies alliance*, White Paper.
Inverardi, P., & Mori, M. (2013). A software lifecycle process to support consistent evolutions. In *Software Engineering for Self-Adaptive Systems II* (pp. 239–264). Berlin, Heidelberg: Springer.
Jumisko-Pyykkö, S., & Vainio, T. (2012). Framing the context of use for mobile HCI. *Social and Organizational Impacts of Emerging Mobile Devices: Evaluating USE: Evaluating Use*, 217–219.
Liang, G., & Cao, J. (2015). Social context-aware middleware: A survey. *Pervasive and Mobile Computing, 17,* 207–219.
Lim, B.Y., & Dey, A.K. (2009, September). Assessing demand for intelligibility in context-aware applications. In *Proceedings of the 11th ACM International Conference on Ubiquitous computing* (pp. 195–204).
Loke, S. (2006). *Context-aware pervasive systems: Architectures for a new breed of applications*. Boca Raton: CRC Press.
Mizouni, R., Matar, M. A., Al Mahmoud, Z., Alzahmi, S., & Salah, A. (2014). A framework for context-aware self-adaptive mobile applications SPL. *Expert Systems with Applications, 41*(16), 7549–7564.
Mori, M. (2011, September). A software lifecycle process for context-aware adaptive systems. In *Proceedings of the 19th ACM SIGSOFT Symposium and the 13th European conference on Foundations of Software Engineering* (pp. 412–415).
Moses, A. (2012). Lg smart fridge tells you what to buy, cook and eat. *The Sydney Morning Herald, January*.
Pascoe, J. (1998, October). Adding generic contextual capabilities to wearable computers. In *Wearable Computers, 1998. Digest of Papers. Second International Symposium on IEEE* (pp. 92–99).
Perera, C., Zaslavsky, A., Christen, P., & Georgakopoulos, D. (2014). Context aware computing for the internet of things: A survey. *IEEE Communications Surveys & Tutorials, 16*(1), 414–454.
Prekop, P., & Burnett, M. (2003). Activities, context and ubiquitous computing. *Computer Communications, 26*(11), 1168–1176.
Rodden, T., Cheverst, K., Davies, K., & Dix, A. (1998, May). Exploiting context in HCI design for mobile systems. In *Workshop on human computer interaction with mobile devices* (pp. 21–22).

Rouvoy, R., Barone, P., Ding, Y., Eliassen, F., Hallsteinsen, S., Lorenzo, J., & Scholz, U. (2009). Music: Middleware support for self-adaptation in ubiquitous and service-oriented environments. In *Software engineering for self-adaptive systems* (pp. 164–182). Berlin, Heidelberg: Springer.

Ryan, N., Pascoe, J., & Morse, D. (1999). Enhanced reality fieldwork: The context aware archaeological assistant. *Bar International Series, 750,* 269–274.

Salehie, M., & Tahvildari, L. (2009). Self-adaptive software: Landscape and research challenges. *ACM Transactions on Autonomous and Adaptive Systems (TAAS), 4*(2), 14.

Schilit, B. N., & Theimer, M. M. (1994). Disseminating active map information to mobile hosts. *Network, IEEE, 8*(5), 22–32.

Schuster, S., Marhl, M., & Höfer, T. (2002). Modelling of simple and complex calcium oscillations. *European Journal of Biochemistry, 269*(5), 1333–1355.

Sumi, Y., Etani, T., Fels, S., Simonet, N., Kobayashi, K., & Mase, K. (1998). C-map: Building a context-aware mobile assistant for exhibition tours. In *Community computing and support systems* (pp. 137–154). Berlin, Heidelberg: Springer.

Taylor, R.N., Medvidovic, N., & Dashofy, E.M. (2009). *Software architecture: Foundations, theory, and practice*. New York: Wiley.

Ur, B., McManus, E., Pak Yong Ho, M., & Littman, M. L. (2014, April). Practical trigger-action programming in the smart home. In *Proceedings of the SIGCHI Conference on Human Factors in Computing Systems* (pp. 803–812). ACM.

Van Bunningen, A. H., Feng, L., & Apers, P. M. (2005, April). Context for ubiquitous data management. In *Ubiquitous Data Management, 2005. UDM 2005. International Workshop on IEEE* (pp. 17–24).

Want, R., Hopper, A., Falcao, V., & Gibbons, J. (1992). The active badge location system. *ACM Transactions on Information Systems (TOIS), 10*(1), 91–102.

Ward, A., Jones, A., & Hopper, A. (1997). A new location technique for the active office. *Personal Communications, IEEE, 4*(5), 42–47.

Winograd, T. (2001). Architectures for context. *Human-Computer Interaction, 16*(2), 401–419.

Zimmermann, A., Lorenz, A., & Oppermann, R. (2007, August). An operational definition of context. In *International and Interdisciplinary Conference on Modeling and Using Context* (pp. 558–571). Berlin, Heidelberg: Springer.

Chapter 3
Elements of Context Awareness

Abstract This chapter describes some essential elements of context-aware applications including context acquisition, context modeling, context reasoning, context distribution, and context adaptation. Context acquisition involves gathering context from the users and the environment. It has to engage with various kinds of sensors. Context modeling which is also called context representation is needed because of the necessary to have the common understanding between the system and its components. It also has to deal with the relationships and dependencies of different types of contexts. Context reasoning involves creating new knowledge and deducing better understanding based on the available context. Context distribution is to deliver the appropriate information or the services to the users. It plays the leading roles, especially for the large-scale network system. Finally, context adaptation involves the adjustment of application's behavior so that the appropriate response can be obtained.

For decades, the research in the context-aware application has been gaining attention dramatically. The expectations come from different requirements of the users from many different application domains as mentioned in Chaps. 1 and 2. However all among those differences, these context-aware applications have the common goal which is to provide the appropriate services to the users and the environment accordingly to their contexts with or without the intervention from the users. More specifically, the context-aware applications are expected to adapt themselves appropriately to satisfy the user and their environment. In summary, the context-aware applications can perform both self-adaptive and self-evolving (Mcheick 2014).

To design and develop efficient context-aware applications, many essential elements are required. Firstly, the context needs to be acquired from various sensors or sources which can be physical and virtual sensors. Then, the collected data needs to be modeled and represented in a meaningful manner. From low-level raw sensor data, the modeled data is proceeded to derive high-level context information. Both contexts are required for the further processes. At the end, the proper functions or services are selected to the users and the environment. As mentioned in Chap. 2, the awareness of applications can be defined as the system that can act rationally and

P. Temdee and R. Prasad, *Context-Aware Communication and Computing: Applications for Smart Environment*, Springer Series in Wireless Technology,
DOI 10.1007/978-3-319-59035-6_3

personally. Although the context adaptation can be seen as the essential element of context-aware applications, some other components are also required. The standard processes from existing applications are mainly considered to identify the key elements of these context-aware systems. For example, Intelligence Cycle (Shulsky and Schmitt 2002) illustrated that the data flow process consists of context collection, context processing, context analysis, context publication, and feedback. WCXMS project (Hynes et al. 2009) illustrated that the data flow process consists of context sensing, context transmission, context acquisition, context classification, context handling, context dissemination, context usage, context deletion, context maintenance, context disposition. Additionally, from the layered architecture proposed by Baldauf et al. (2007), the functions necessary for the context-aware system are context sensing, context retrieval, context preprocessing, and context storage. Context lifecycle (Perera et al. 2014) normally consists of context acquisition, context modeling, context reasoning, and context dissemination. From these examples, it can be concluded that the principal elements of context-aware applications may include five main elements including context acquisition, context modeling, context reasoning, context dissemination and context adaptation as they can be found as the common elements of the existing context-aware applications.

This chapter thus will discuss five essential elements for being context awareness including context acquisition, context modeling, context reasoning, context dissemination, and context adaptation. Context acquisition mainly focuses on how to gather different types of contexts from heterogeneous sensors. Context modeling mainly focuses on how to represent context in the form that can be understood by machines. Context reasoning mainly focuses on how to interpret or make sense of the low-level context. Context dissemination mainly focuses on how to represent the context or the service to the users or their environment. Lastly, context adaptation mainly focuses on how to select the appropriate functions or services to the users and the enviornment. Along with the demonstration of the existing works, limitations and some significant concerns are also suggested in this chapter.

3.1 Context Acquisition

Context acquisition is the first element required for context-aware applications. It involves gathering context from the users and the environment. It has to engage with various kinds of sensors such as physical sensors, virtual sensors, and logical sensors. Because of the advance in computing and communication technologies, the sensors now are smaller, distributed and even embedded in the objects of our daily life (Schilit et al. 1994; Schmidt et al. 2002; Olifer and Olifer 2005; Yamabe et al. 2005). The context-aware applications will not properly work if they cannot perceive the correct real world information by their sensors. This section aims to show the common concerns (Perera et al. 2014) accordingly to context acquisition including responsibility, event frequency, context source, sensor types, and acquisition process.

3.1.1 Responsibility

Generally, the context can be acquired by using two methods including push and pull methods (Pietschmann et al. 2008). Pushing method is responsible for obtaining sensor data periodically from both physical and virtual sensors. The physical sensors make the major decisions by itself on sensing and communicating. There is typically less sent information for decision making, and the sensors need to be re-programmed when the requirements are changed. This kind of method can be used in any application having the sensors with enough knowledge and power to perform context reasoning locally. Generally speaking, this approach is suitable for any application when the event can be detected by some sensors. Moreover, the sensors do not need any software to perform reasoning and evaluating the conditions of their environment. On the other hand, pulling method requires the software for sensing and communicating sensor data. The software also makes the decision on when to collect the data. More communication bandwidth is needed for sending the requested data to the sensors periodically. This method can be used in any application when the sensors do not have knowledge about reasoning and want to send the data for making a decision. This approach is suitable for any application when the event is detected by collecting, processing, and reasoning a large amount of sensor data.

3.1.2 Event Frequency

Two different event types can be found in context acquisition including instant and interval events. Instant or threshold violation events mean that the events occur instantly and the sensor data needs to be acquired immediately when the event occurs. Both push and pull methods can be used for obtaining this kind of event. The examples of this event are opening the window, turning on the air conditioner, etc. Since the data will be gathered as soon as the conditions are met, more knowledge is required to identify and satisfy the condition. The sensors should know what exact they want. It's hard to detect the events that require various types of data from heterogeneous sensors. Consequently, this method potentially consumes more energy for data processing. In summary, the applications which are appropriate for this kind of event can be any application where the expected outcome is well-known either by hardware or software levels. The examples are the heat detection for agricultural production, person detection for smart home application, etc.

Interval events are the events that span through a given period. The sensor data needs to be acquired periodically. Both push and pull methods can also be used to collect this event. The sensors do not need to either be intelligent or have outstanding capabilities for processing and reasoning. However, the reasoning method requires the software to deal with information changing over time.

Additionally, this event may cause wasting the energy due to the redundant of data communication. The applications applicable for this event type are the application having the situations that either hardware or software sensors do not know the expected outcome of the application. The well-known application is air pollution monitoring system that the temperature and Carbon dioxide gas are periodically measured.

3.1.3 Context Source

The context acquisition directly involves the context sources which can be classified by the origin where the context comes from including directly from the sensors, through a middleware infrastructure, and from the context server (Chen et al. 2004).

Sensor directed acquisition gathers data from the hardware sensors attached locally or related Application Programming Interfaces (APIs) through any kinds of communication. Most sensors require software driver support. Moreover, the software drivers and the libraries need to be installed locally for some sensors. However, the current wireless technology can allow data transmission without the local installation of the driver. This method is efficient as it allows direct communication with the sensors. The advantage is that there is the control over sensor configuration and data retrieval process. The significant technical knowledge is required such as programming and configuring in hardware level. However, the dedicated time, effort and cost are usually major concerns. Additionally, updating the sensor is also tough because of the customization between the hardware sensor and the application. This acquisition technique works correctly for small scale scientific experiment and for the situation where the limited numbers of sensors are involved.

Through middleware acquisition gathers sensor data from middleware solution. This method is easy to manage and manipulate context because the middleware is designed to take the responsibility of management tasks. It can manage data faster with less effort and technical knowledge. It is also easy for customization of hardware sensors and application. However, this method requires more resources and has less control over sensor configuration. The applications suitable for this approach are IoT applications having a large number of various sensors.

Context server acquisition gathers the context from several other context storages such as the database, web services, etc. This method is useful when there is the limitation of computing resources. Moreover, this method requires fewer resources and can manage data faster with less effort and knowledge. However, there is no control over sensor configuration. The applications suitable for this method are frequently used in the situations where there is the limitation of resources while a significant amount of context is required.

3.1.4 Sensor Type

The sensor can be considered as a mean of measuring physical entities. The obtained data can be used by a system or application to determine appropriate actions. The combination of multiple sensors can give more detailed data for the system or the application. Nowadays, a large variety of sensors are able to sense different kinds of information from almost every measurable entities such as temperature sensors, touch sensors, motion sensors, pressure sensors, light sensors, etc. With the advance of sensing technology, the sensors are more attractive and affordable nowadays. For example, positioning technologies and short-range networking technology have been gaining much interest worldwide (Hightower and Borriello 2001) for obtaining location information rather than the satellite networks. Many different types of sensors can detect the position of entities such as GPS, the Radio Frequency Identification (RFID), etc. Especially for RFID tag, it is also known as a smart label (Lahiri 2005) that can be read from and written to by using a RFID reader which uses the energy from radio frequency field. RFID tags can store from 64 up to several thousand bits of data. Although the location is widely used in many context-aware applications, the important concerns are what sensors should be used and where the sensors should be placed appropriately for each application.

Nowadays, the sensors can be embedded in the environment as part of the house, the office, the car, worn on people or even placed within people. Thus, networks of sensors have been playing the main role for many applications (Zhao and Guibas 2004) such as warehouse inventory management, automotive applications, environmental monitoring, military and security, line production, etc. Such sensor networks comprise numbers of sensors scattered over particular areas and they are configured to transmit information at an appropriate rate within the specified predefined time duration. Many context-aware applications have utilized the use of context to recognize everyday situations such as the prediction and identification of human interrupt ability (Fogarty et al. 2005; Ho and Intille 2005), the determination of mobile phone location (Gellersen et al. 2002), etc. In conclusion, it can be seen that the sensors can be employed in the common setting through variety kinds of applications. Current studies are paying much interest in the sensors that can be worn unobtrusively and could proliferate in the environment.

For general way of classification, the sensors can be divided into three categories (Indulska and Sutton 2003) including physical, virtual and logical sensors. Physical sensor is the sensor that can generate the data by itself. The retrieved data is just called low-level context. It can be considered as less meaningful and sensitive to the small changes. It is used for collecting observable physical phenomenon such as light, temperature, humidity, etc. On the other hand, the virtual sensors obtain the data from many sources and publish it as sensor data such as calendar, contact list, email, etc. More importantly, these sensors do not generate the data by themselve and do not physically appear. Therefore, it is often used to collect the information that cannot be measured physically. Finally, a logical sensor is also called as the software sensor. It combines physical sensors with virtual sensors to produce more meaningful information. However, these types of sensors do not have the control over data production process. Therefore, they can be used to collect information that is impossible to receive directly through either single physical sensor or single virtual sensor. This type of sensor requires the complex and costly processing and fusing methods.

3.1.5 Acquisition Process

There are general three ways to acquire context (Alegre et al. 2016) including sensing, deriving and manually providing methods. Sensing method gathers the data through the sensors. Deriving method obtain the information by performing any computational operation on the collected sensor data. Finally, for the manually provinding method, the user provides context information manually via predefined setting operations.

As mentioned before that the low-level context can be used for constructing higher-level context, the common way is to employ context abstraction service which is shown in Fig. 3.1. The common components for this service include widget, interpreter, and aggregator. The widget which is the software module is used to gather the context data from the physical sensors directly. Basic operation can be performed here such as feature selection, data fusion, subscription, etc. The widget can represent current value, history, and subscriptions, etc. Therefore, the low-level context can be abstracted to be the higher level context in some degree. More abstract context can be derived by using the interpreter. The interpreter, which can be any data processing and analyzing method, will be used by widgets, aggregators, and applications. Finally, the aggregator will select relevant context and allows the context abstraction to provide appropriate higher level context for the applications. The significant concern for acquisition process is to choose the suitable and efficient components for different applications.

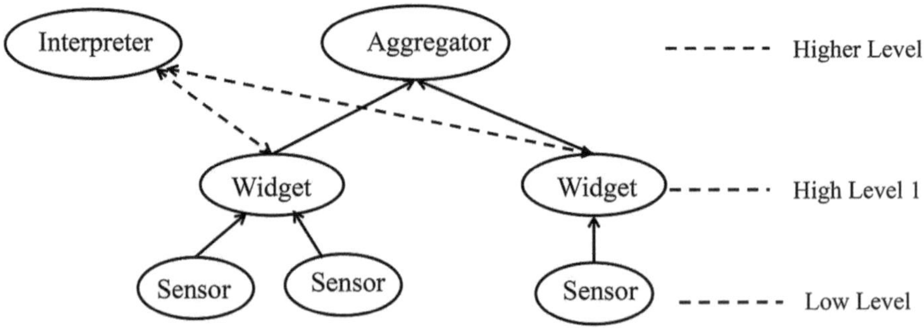

Fig. 3.1 Context abstraction service

3.2 Context Modeling

Since the overall goal of the context-aware application is to develop the applications that are evolvable. A good context modeling will reduce the complexity of the application while maintaining the ability to adapt or evolve (Bettini et al. 2010). Context modeling is needed because of the necessary to have the common understanding of the system and all components. It is also useful for reusing and sharing of context information among applications because gathering and maintaining context are typically expensive. The formal representation of context is also necessary for consistency checking and sound reasoning. Additionally, the well-designed modeling technique will make the development much easier.

Context modeling is also called context representation. Over the last decades, many context modeling has been developed varying from the simple models to complicate model to promote the heterogeneity and the mobility. Context models normally have to deal with a large variety of context sources with the different acquisition rate. For example, physical sensors sense the real world and provide nearly real-time access, and the context needs to be interpreted before being used by the applications. The virtual sensors are rarely updated and typically do not require further interpretation. New context can be derived from the existing context. The context from the database is usually static differing from the context gathered from mobile devices. Consequently, there is necessary to have the context model that can deal with heterogeneity and mobility of the context sources. Because of heterogeneity, it is possible that the gathered context will not be perfect. Therefore, the modeling needs to deal with imperfection. Moreover, some contexts may be the conflict with other contexts. Also, the modeling techniques need to address the fault of context effectively.

The context modeling also has to deal with the relationships and dependencies of different types of contexts. There are a large variety of context types for capturing the real world behavior. Moreover, one context may depend on other contexts. Additionally, context modeling needs to deal with timeliness. It is more often that the applications need to access the historical state and the future state of the contexts. Especially it is not easy to manage the historical contexts for some applications where it may not be feasible to store every context value for the future access. Therefore the employment of particular modeling techniques is required. Selecting the modeling techniques need to also concern about the reasoning technique used for applications. Context reasoning is used to derive the new context to be higher level contexts that can model the real word more precisely. The detail of context reasoning will be discussed in the next section.

There are several context modeling techniques (Balavalad et al. 2009; Baldauf et al. 2007) depending on the model of context. Context model can be static or dynamic. While the static models have a predefined set of context information that will be collected, the dynamic models do not have the predefined set of context information before (Yanwei et al. 2011). Typically, there are two steps for representing context. Firstly, new context needs to be defined regarding many concerns such as attributes, qualities, property, characteristics, relationships, etc. Then, the result of context modeling needs to be validated, merged and added to the existing repository for the further usages. Currently, there is no standard to specify context modeling. This task can be considered as a subjective decision. At the high-level, the implementation of these techniques can vary depending on the application domain. All context modeling techniques (Chen and Kotz 2000; Strang and Linnhoff-Popien 2004) always have their strengths and weaknesses. Therefore, they will be selected regarding the application. For example, MoCA (Elnahrawy and Nath 2004) uses an object-oriented approach to the model context by using XML. The context models normally consist of structural information, behavioral information, and context-specific abstraction. The structural information includes attributes and dependencies among context types. The behavioral information includes the context attributes having a constant or a variable value. The context-specific abstraction consists of the contextual events and the queries. W4 Diary (Castelli et al. 2009) uses a W4 (who, what, where, when) based context model to structure data for extracting high-level information from the location data. Nowadays, there are many well-known context modeling techniques such as key-value, markup schemes, graphical based, object-based, logic based, and ontology-based modeling (Perera et al. 2014). The detail of these modeling techniques is briefly described in this section.

3.2.1 Key-Value Modeling

This method is the simplest form of context representation among all other techniques. It uses the key-values pairs to model context information in different formats such as text files and binary files. It is easier to be used for the small amount of data, but it is not scalable and not suitable to keep the complex data structure. This technique cannot model hierarchical or relationship structures. It contains mostly independent and non-related information, which is suitable for less complex and temporary modeling requirements with limited data transferring (Bettini et al. 2010). Therefore, it is an application oriented suiting the temporary storage as less sophisticated applications of configurations and user preferences. Although this technique is easy to develop, some concerns have to be taken into account. The main concerns of this technique are their limitations in capturing a variety of context types, dependencies, and timeliness. Moreover, there are also some minor concerns for this technique such as quality controlling, consistency checking, uncertainty ensuring, and suitable context reasoning.

3.2.2 Markup Scheme Modeling

This modeling technique models data by using tags. It is the improvement over the key-value modeling technique by allowing more efficient data retrieval. The validation tool as XML is available for popular markup technique. Consequently, XML is widely used in almost all application domains to store and transfer data among applications and their components. However, this technique does not allow reasoning. Therefore, the interoperability and reusability over different markup schemes can be difficult without the appropriate design. A typical application of this technique is profile modeling which is commonly developed with XML. However, other languages supporting tag based storage can also be found to support markup scheme modeling such as JavaScript Object Notation (JSON), Tuples (Yanwei et al. 2011), etc. The Composite Capabilities/Preference Profiles (CC/PP) (Nilsson et al. 2000) can also be considered as one of the popular markup scheme modeling and the first modeling technique that uses Resource Description Framework (RDF) which is one of the well-known standards for semantic technology.

3.2.3 Graphical Modeling

This modeling technique models context with relationships. The popular tools are Unified Modeling Language (UML) (Rumbaugh et al. 2004) and Object Role Modeling (ORM) (Halpin 2001). This modeling technique allows the relationship to be captured by the context model. Therefore it has more efficiency than markup and key-value techniques if there are enough required resources. It can be found that this technique is easy to learn and to use. Historical context can also be stored in databases that can contain massive amounts of data and provide simple data retrieval operations. However, the different implementation may cause it to be not easy to achieve interoperability. The requirement of context retrieval may require complex Structured Query Language (SQL) queries. Therefore, adding more context information and changing the data structure can be very challenging.

3.2.4 Object Based Modeling

This technique makes use of object-oriented concepts by using the encapsulation and the inheritance to represent context in the form of programming code level. It aims to model data by using class hierarchies and their relationships. Therefore, this modeling technique can promote encapsulation and reusability. Moreover, this technique can be easily integrated into existing context-aware applications because most of the programming languages support object-oriented concepts. The object-based modeling is suitable for an internal, code based, run-time modeling. Hydrogen project (Hofer et al. 2003) is the popular system using object-based modeling.

3.2.5 Logic Based Modeling

This technique uses the logical based method to define the formal model. More specifically, the facts, expressions, and rules are mainly used for constructing knowledge. At the same time, rules are mostly used to describe required entities such as policies, constraints, and preferences. Different facts can be inferred separately. It can be seen that the low-level context is used for generating high-level one. More specifically, the higher-level context can be derived with the existing rules and knowledge. Additionally, reasoning is needed for the creation of higher-level context. Therefore, logic-based modeling is usually used for modeling events and actions to suit the reasoning process. Many existing concrete structures and languages can be used for this kind of modeling. Furthermore, interactive techniques can be employed for non-technical users to develop logic based or rule based representations quickly. Although its capability is likely to be higher than other context modeling techniques, the reusability and applicability frequently decrease because of the lacking of standardization.

3.2.6 Ontology Based Modeling

This technique describes taxonomies of concepts and relationships. Therefore, the context represented by ontologies uses several semantic technologies. Some standards are such as Resource Description Framework (RDF) (Pan 2009), Web Ontology Language (OWL) (Bechhofer 2009), etc. Ontologies use a particular language to represent the context and its relationship. A full range of development tools and reasoning engines are also available. There are many reasons that many context-aware applications employ ontology-based modeling technique (Wang et al. 2004; Noy and McGuinness 2001). For example, the ontology is useful to share a common understanding of the structure of any entities. It is also useful for analyzing of domain knowledge by separating from operational knowledge. The domain knowledge can be reused appropriately. The higher level knowledge is easy to be inferred. The ontology based technique is also able to make domain assumptions to be more explicit. According to many surveys of context-aware applications, the ontologies seem to be the preferred mechanism of modeling context for academics and industries. However, context retrieval from this modeling method can be computationally intensive and time consumption especially when the number of data increases.

From pieces of literature, no single modeling technique can be used ideally as standalone method. Multiple modeling techniques are recommended as the best way to provide the practical context-aware applications. Moreover, there is also the strong connection between modeling techniques and reasoning techniques. Next section will introduce context reasoning techniques in detail.

3.3 Context Reasoning

After being modeled, the sensed context will be used for obtaining new knowledge. This task is called context reasoning. Context reasoning can be defined as a method of creating new knowledge and deducing better understanding based on the available context (Bikakis et al. 2007). It can also be considered as a process of deducing high-level context from a set of low-level contexts (Guan et al. 2007) or so-called inference. Like context modeling, the reasoning techniques need to deal with uncertainty and context imperfection such as ambiguous, imprecise, or erroneous, etc. The important performances of reasoning performance are such as efficiency, completeness, interoperability, soundness, etc.

For reasoning, context information has to be manipulated through four primary processes (Nurmi and Floréen 2004) including pre-processing, data fusion, inference, and reasoning. Context pre-processing is required because the contexts are frequently gathered incompletely from hardware sensors. Together with network communication, collected data may not be entirely collected. Therefore, the received data needs to be cleaned. Several cleaning methods are used in this process such as filling missing values, removing outliers, validating context via multiple sources, etc. Next, data fusion is performed to obtain more accurate and complete data that cannot be achieved by using a single sensor. The data fusion can be done by combining sensor data from multiple sensors (Llinas and Hall 1998). Next, context inference can be made from either a single interaction or multiple interactions. For example, W4 Diary (Castelli et al. 2009) representes the context as Who is doing What, Where and When. The low-level context can be inferred from some reasoning mechanisms to generate more meaningful results. For example, someone is walking in the shopping mall in Bangkok at 17.00 pm. For the first iteration, the coordination information of a GPS sensor may be inferred as one particular shopping mall in Bangkok. For the next iteration, one particular shopping mall in Bangkok may be inferred as someone's favorite mall in Bangkok. Every iteration can give more meaningful information. Context reasoning techniques can be classified broadly into six categories (Perera et al. 2014) which are originated and are typically employed in the fields of artificial intelligence and machine learning including rules, fuzzy logic, supervised learning, unsupervised learning, ontological reasoning, and probabilistic reasoning. This section describes more detail of each reasoning method.

3.3.1 Supervised Learning

Supervised learning is one type of learning algorithms for machine learning. For supervised learning, training examples are collected and labeled according to the expected targets or results. Then, a function that can generate the desired results using the training data is derived which is called generalization function.

Generalization refers to the ability to give reasonable outputs for any inputs that are not trained during the training process. This technique is usually fast and accurate. However, the efficiency depends really on the training process. For context-aware applications, this technique is widely used in mobile phone sensing (Lane et al. 2010) and activity recognition (Riboni and Bettini 2009). The example techniques are decision tree, Bayesian networks, artificial neural network, support vector machines, etc.

3.3.1.1 Decision Tree

A decision tree is a supervised learning technique where the tree structure is built from a dataset for classifying data. It is a predictive model where the leaf represents a classification. Each branch represents a conjunction of features causing the target classification. The main advantage of a decision tree is that the classification rules that are easy to understand and explain can be generated from the given data set. For context-aware applications, these rules are also useful in analyzing sensor performances and feature extraction (Bao and Intille 2004). The efficiency of this technique is highly dependent on the size of training data. Therefore, for a large real-world data set, the efficiency is still the challenging issue.

Figure 3.2 demonstrates the example of using decision tree to perform simple classification task for buying new smartphone problem. From the figure, the node represents the attributes for splitting the data, the branch represents the class of attribute, and the leaf node represents the decision. For this example, the attributes are the age range, the student status and the income range. The branches represent classes of each attribute. Age range has three classes including <=30, 31–40, and >=40. The student status has two classes including yes and no. Lastly, the income range has two classes including high and low. After analyzing the collected data, the decision tree can be constructed with several methods. The popular one is

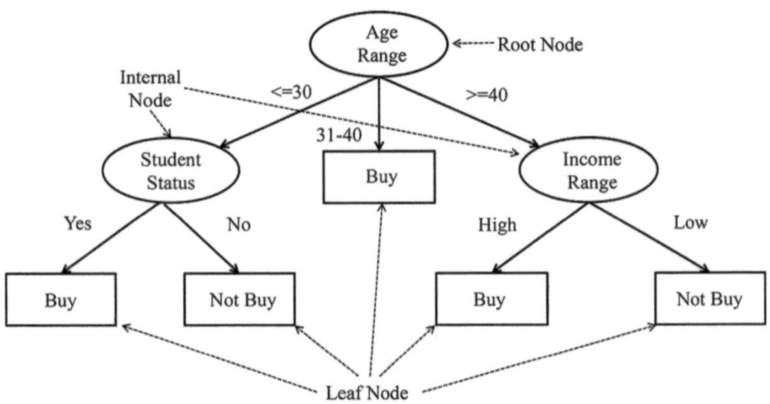

Fig. 3.2 Decision tree for buying new smart phone

using entropy and information gain for calculating the homogeneity of the data as in ID3 algorithm. More specifically, the entropy and the information gain are used for determining the best node for spliting the data at each step. From this example, the Age range is the root node while the student status and the income range are considered as internal nodes. The decisions are shown as the leaf nodes which are buying or not buying. The rules are finally constructed for classifying or grouping the testing data to categorize them into the right class. Generally speaking, the objective of those rules is to make the decision for the new testing cases. From this example, some decision rules are if the person is between 31 and 40, then this person will buy the new smartphone, if the person is more than 40 and has low income range, then this person will not buy the new smartphone, etc. Decision tree reasoning technique is widely used by context-aware applications. The famous examples are activity recognition (Mathie et al. 2004) and student assessment system (Huang et al. 2008).

3.3.1.2 Bayesian Networks

Bayesian networks employ to explain the uncertainty of data by deriving conditional probability from Bayes's Theorem. The derivation of Bayes' Theorem is shown in (3.1).

$$P(A|B) = \frac{P(B|A)P(A)}{P(B)} \tag{3.1}$$

where $P(A|B)$ is known as "posterior probability" or the probability of A after considering the effect of B, $P(B|A)$ is called the likelihood, $P(A)$ is prior probability of A, $P(B)$ is probability of B and can be regarded as scaling factor.

Bayesian networks, which are so-called belief networks, are reasoning method under uncertainty condition in the form of graphical models. The nodes represent variables that can be discrete or continuous. The arcs or branches represent direct connections between them. All branches are the directed graph that means they point to the particular direction. No cycle means that there will not be the branch leading back to the starting node. All nodes in a Bayesian network represent a set of random variables, which is $X = X_1, \ldots, X_i, \ldots, X_n$ from the domain. A set of directed arcs or links or branches connects pairs of nodes, $X_i \rightarrow X_j$, representing the direct dependency relation between those variables. For discrete variables, the strength of the relationship between variables is quantified by the conditional probability distributions which are associated with each node. At the same time, missing arcs implies conditional independence. The full joint distribution is defined as the product of the local conditional distributions as shown in (3.2).

$$P(X_1, X_2, X_3, \ldots, X_n) = \prod P(X_i|parents(X_i)) \tag{3.2}$$

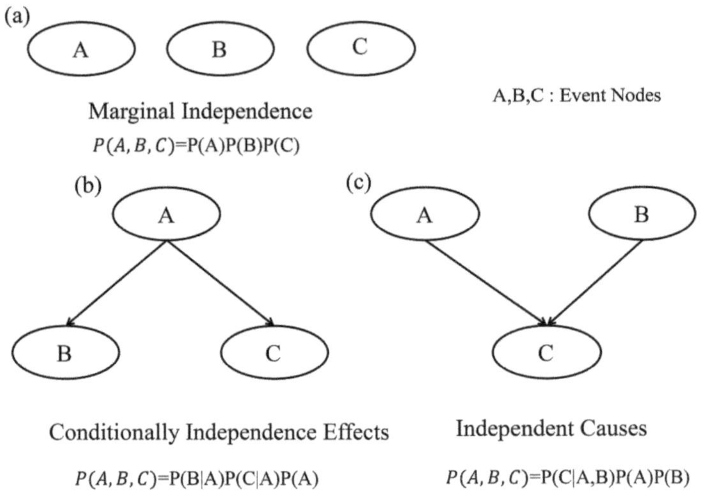

Fig. 3.3 Conditional independence structures

where $P(X_1, X_2, X_3, \ldots, X_n)$ is the joint distribution, and $P(X_i|parents(X_i))$ is the local conditional distributions.

Simple example Bayesian network is shown in Fig. 3.3 which the graph structure represents conditional dependent relations of nodes. A, B, and C are the event nodes. From Fig. 3.3, there are three possible structures to represent conditional independence among event nodes. From Fig. 3.3a, all nodes are not connected, so they are marginal independence, which can be defined as (3.3). From Fig. 3.3b, B and C are conditionally independent given A, so the conditionally independent effects can be defined as (3.4). For example, if A is a disease and B are conditionally independent symptoms given A disease. For Fig. 3.3c, A and B are marginally independence but become dependent once C is known. For example, Given C, observation A makes B less likely. Therefore, the independent causes can be seen as shown in (3.5). The structures are chosen to explain each problem appropriately.

$$P(A, B, C) = P(A)P(B)P(C) \tag{3.3}$$

$$P(A, B, C) = P(B|A)P(C|A)P(A) \tag{3.4}$$

$$P(A, B, C) = P(C|A, B)P(A)P(B) \tag{3.5}$$

For reasoning, Bayesian networks describe conditional independence among the subsets of variables allowing the combination of prior knowledge about dependencies among variables from the training data. Bayesian networks are commonly used in combining uncertain information from a large number of sources and deducing higher-level contexts (Ko and Sim 2008; Park et al. 2011).

3.3.1.3 Artificial Neural Networks

Having the inspiration by biological neuron networks, Artificial Neural Networks (ANNs) are widely accepted in many application domains. They are good in automatically learning of the complex problem by mapping and extracting a non-linear combination of the set of features. A neural network is composed of many artificial neurons that are linked together according to specific network architecture. More specifically, it consists of a large number of highly interconnected processing elements, which is called neurons, works together. Most ANNs have some training algorithms that they can learn from the examples and exhibit some degree of generalization out of the training data. During the training period, ANNs adapt itself by using some examples of similar problems with and without the desired solution. After sufficient training, the trained system can provide the solution relating to inputs to outputs. Moreover, it can offer an alternative solution to the new problem.

Figure 3.4 shows the structure of ANNs consisting of an input layer as $x_1, x_2, x_3, \ldots, x_n$, a hidden layer, and an output layer $y_1, y_2, y_3, \ldots, y_m$. The mapping between inputs and outputs are performed at hidden layer with the composition of activate functions f. The weights between input and hidden layers are w_{ij} and the weight between hidden layer and output layer are v_{jk}. The total input y_{in} is defined as the summation of all inputs multiply with their associated weights. The biasing node can be used. The total input of ANNs is shown in (3.6).

$$y_{in} = b + \sum_{i,j=1,1}^{n,m} x_i w_{ij} \tag{3.6}$$

The output of each hidden node or y depends on the activation functions used as shown in (3.7). The popular activation function is Sigmoid function which is shown in (3.8).

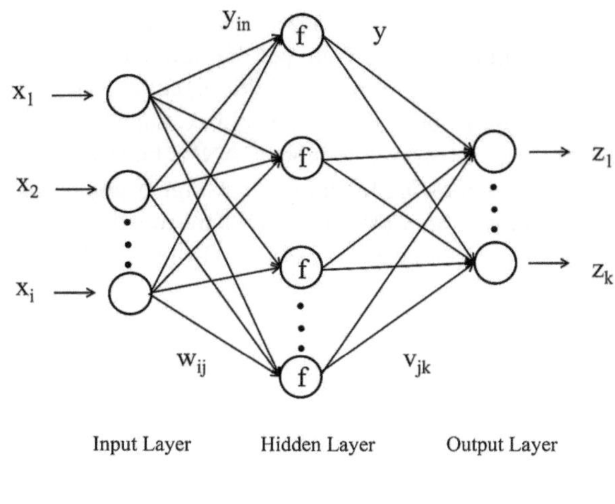

Fig. 3.4 Structure of artificial neural networks

$$y = f(y_{in}) \tag{3.7}$$

$$f(x) = \frac{1}{1 + \exp(-x)} \tag{3.8}$$

The output at output layer z_{in} is also defined as the summation of all inputs multiply with their associated weights as shown in (3.9). The output at output layer is also the function of z_{in} as shown in (3.10).

$$z_{in} = \sum_{j,k=1,1}^{m,j} y_j v_{jk} \tag{3.9}$$

$$z = f(z_{in}) \tag{3.10}$$

ANNs perform learning by modifying the weights. Every weight is modified by specific learning rules. For example, delta rule (Hagan et al. 1996) which is often used by the common class of ANNs called Backpropagation Neural Networks modifying the weight by using learning rate numbers (α) ranging between 0.1 and 1. The changing of new weight from the old weight when the difference between the actual output (t) and the target output (T) as well as the input (I) at each layer are considered is shown in (3.11). Generally, the initial weights are randomly chosen between −1.0 and 1.0 or −0.5 and 0.5.

$$w_{new} = w_{old} + \alpha(T - t)I \tag{3.11}$$

From Fig. 3.4, the raw information is fed into the network through the input units. At each hidden node, the activity is determined by the input nodes and their associated weights (W_{ij}). At the output layer, the activity is determined by the hidden units and their associated weights (V_{jk}). The hidden units are selected freely accordingly to their representations of the input. The weights between the input and hidden units are modified for every iteration of the training process. Finally, the knowledge is obtained in the values of the connection weights. ANN's performance is highly dependent on the amount of training data and the numbers of training. It is considered as a good choice if there is plenty of training data. The suitable problem for ANN is usually poorly understood to derive the approximate model. Additionally, the noisy data also affect the performance of ANNs. ANNs have widely been used in many application domains such as intelligent control, signal processing, pattern recognition, etc. For context-aware applications, ANNs are mostly used for activity recognition (Favela et al. 2007), healthcare monitoring (Korel and Koo 2010), etc.

3.3.1.4 Support Vector Machines

Support Vector Machines (SVMs) are well-known as the methods using both linear and nonlinear mappings for classifying data by transforming the training data into the higher dimension. Within this dimension, there is the hyper-plane separating the training data of one class from another class. The data can be divided into two classes by using high dimension together with suitable nonlinear mapping. SVMs can handle large data by employing over-fitting protection. The ability to separate from binary classification can be enhanced by maximizing the margin according to the intuition as shown in Probability Approximately Correct (PAC) theory (Haussler 1990). The primary concern for linear separation is which decision boundaries can separate the most optimal for two classes. The best boundary should be far away from the data of both classes as much as possible while it should be as close as enough to provide the proper separation. However, SVM is sensitive to the noise data. Consequently, a relatively small number of mislabeled examples can dramatically decrease the classification performance of SVM. Additionally, SVM can consider only two classes. Performing multi-class classification with SVM requires multiple SVM's outputs. Predicting output from new inputs can be done by comparing the furthest distance into the positive region of each SVM. Figure 3.5 shows the concept diagram of SVM for separating two classes.

Figure 3.5a shows that there are many decision boundaries for this separation problem. However, it is difficult to determine the most optimal one. Instead of having exact boundary, SVMs have the margin that can enhance the ability of separation as shown in Fig. 3.5b. Margin ρ of the separator is the distance between support vectors and r is the distance from example x_i to the separator and can be shown in (3.12), where w the weight vector and b is the biasing value. For linear

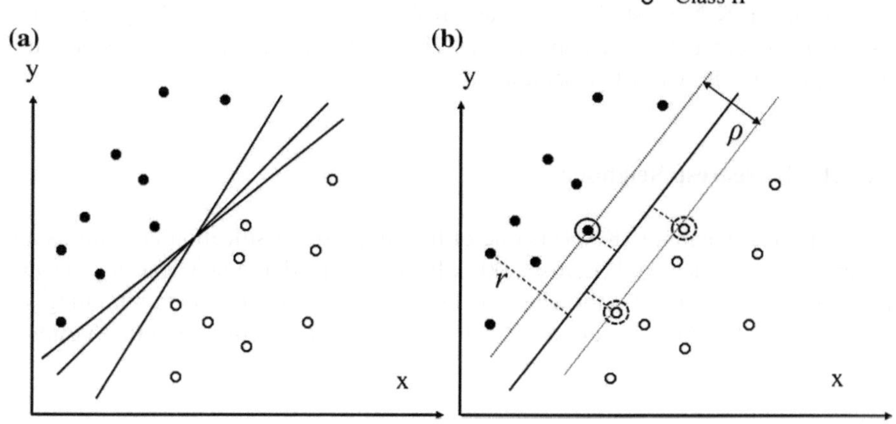

Fig. 3.5 Classification concept for linear separator and SVM

SVMs, the margin ρ can be defined by (3.13), when x^{+}, x^{-} are the data from both two classes. The margin has to be the maximum one.

$$r = \frac{w^T x_i + b}{\|w\|} \tag{3.12}$$

$$\rho = \frac{(x^{+} - x^{-}) \cdot w}{|w|} = \frac{2}{w} \tag{3.13}$$

SVMs have been used successfully in many real-world problems especially in classification problems such as text categorization, image classification, hand-written character recognition, bioinformatics for protein and cancer classifications, etc. For context-aware applications, it also has been widely used to detect activity recognition (Patel et al. 2007) of patients in the public health domain (Doukas et al. 2007) and in a smart home environment (Reignier et al. 2009).

3.3.2 Unsupervised Learning

While the supervised learning trains the collected examples and labels them according to the expected results, the unsupervised learning is not provided with the expected results during the training process. Due to no training data, there is no error or reward signal to evaluate a potential solution. Some intrinsic structures are found all among data after the data is explored. Clustering can be considered as the most common form of unsupervised learning. It is the process of grouping a set of similar objects into the same class which is called the cluster. More specifically, clustering is a technique for finding similarity groups in data. It consolidates similar data instances or near each other into one cluster while the different data instances or far away from each other into different clusters. Unsupervised learning techniques such as K-Nearest Neighbour and Kohonen Self-Organizing Map are widely used in many context-aware applications. The details of these unsupervised learning techniques are shown in this section.

3.3.2.1 K-Nearest Neighbor

The K-nearest neighbor (KNN) is one of the simplest classification algorithms and is very easy to understand (Larose 2005). It can be called as a lazy learning because the learning will happen only when the test example is given. Given a training set, KNN can predict the class of an unseen instance by comparing it to other points in the space. For KNN, the number "K" is a user-defined parameter to determine the number of clusters, and it is a positive integer. Both low and high values of K have their advantages. However, it is not easy to determine for some particular problems. The best value of K depends on the data. Cross-validation is used to compare efficiencies. The classes with more examples tend to dominate the predictions of unknown instances. Although it is the straightforward algorithm and it is easy to employ, but it can be computationally intensive depending on the size of the training set.

KNN starts with remembering all training examples. Given the new example or testing data x, KNN find the closest distances of training examples (x_i, y_i) and then predict the class of y_i. Normally, there are many types of distances used to measure the distance between unknown new data and those in the training example. The examples are such as Euclidean Distance, Minkowski Distance, Mahalanobis Distance, etc. The Euclidean distance is found frequently. More specifically, the Euclidean distance between sample x_m and x_n is defined as in (3.14). The smaller Euclidean distance means that two examples are more likely to be the same class.

$$d(x_m, x_n) = \sqrt{\sum_i (x_{m,i} - x_{n,i})^2} \tag{3.14}$$

The principle of KNN is shown in Fig. 3.6. From the figure, the new example is classified as Class B when K = 4. Since three examples of class B are the closets and only one example of class A is the closet, the majority vote is given that the new example belongs to class B. For K = 3, the majority vote also classifies this new example to class B. When K = 2, the classification cannot be done in this example. Therefore, the number K is important for KNN and requires carefully selection.

Fig. 3.6 KNN principle concept

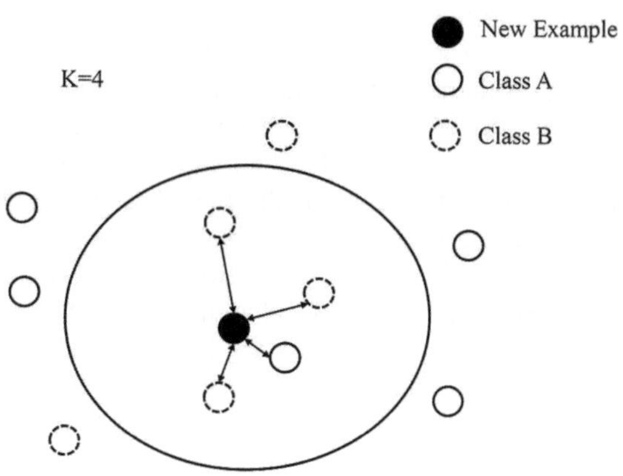

Fig. 3.7 Principle of SOM

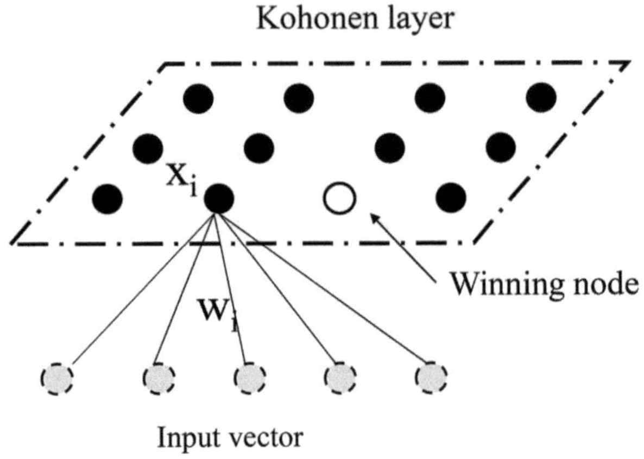

KNNs can be considered as the non-parametric techniques, and they have been widely used for solving the problems in statistical estimation and pattern recognition at the beginning of 1970s. For context-aware applications, KNN is used in the classification of low-level context or from sensor hardware levels high-level tasks such as indoor and outdoor positioning and location (Lin and Lin 2005).

3.3.2.2 Kohonen Self-Organizing Map

Kohonen Self-Organizing Map or Self-Organizing Map (SOM) is one type of artificial neural network having unsupervised learning algorithm (Kohonen et al. 2001). SOM is also called Topology Preserving Maps showing the idea of the clustering neighbor unit. During the self-organizing process, the weight vectors of winning unit and its neighbors are updated. SOM is named as self-organizing because no supervision is required. The word "maps" is assigned because SOM attempts to map its weights to conform the given input data. Every node in a SOM network can learn by trying to become like the input nodes having the data presented to them. The structure of SOM is shown in Fig. 3.7.

After the weights of each node are initialized, a vector is chosen randomly and presented to the network. Then, the calculation is performed to examine which node that its weights are most like the input vector. The winning node is called as the Best Matching Unit (BMU) which is typically determined by the Euclidean distance between input and weight vector and shown in (3.15). The radius of the neighborhood of the BMU is calculated and set to be the radius of the network. Any nodes within the radius of the BMU are adjusted for making them more like the input vector. The closer node to the BMU represents that its weights are more adjusted. The learning process typically takes several iterations.

$$d(x, w) = \sqrt{\sum_i (x_i - w_i)^2} \tag{3.15}$$

where x is the input vector and w is the weight vector. The node having minimum distance is considered to the BMU or the winning node. Its weights are adjusted as shown in (3.16).

$$w_{i,j}(new) = w_{i,j}(old) + \alpha\big(x_i - w_{i,j}(old)\big) \tag{3.16}$$

where $w_{i,j}(new)$ is new weight, $w_{i,j}(old)$ is old weight, and α is the learning rate having the value between 0.1 and 1.

SOMs are widely used in image classification applications. For context-aware applications, SOMs are used in many classification problems such as classifying the incoming sensor data in a real-time fashion (Van Laerhoven 2001), capturing user contexts by dynamic profiling (Shtykh and Jin 2008; Korel and Koo 2010), etc.

3.3.3 Rule Based Method

This method is the most straightforward and traditional way of reasoning. Rules are usually structured in an IF-THEN-ELSE format for generating of high-level context information from low-level context. Although it is easy to develop, this method usually has the limitation of generalization and numbers of rules. This approach has been combined with other methods recently to promote more efficient generalization such as ontological reasoning (Keßler et al. 2009) for event detection

Table 3.1 Higher level context generation with rule based method

Situation	Reasoning rules
Cooking	Located (Kitchen) ∧ Moving (Yes) ∧ Oven status (On) ∧ Room light (On)
Outdoor	Temperature (Cold) ∧ Humidity (Humid) ∧ Light type (Natural) ∧ Light intensity (Bright)
In Theater	Located (Indoor) ∧ Room light (Dark) ∧ Sound condition (Noisy) ∧ Moving (No)

(Barbero et al. 2011). The examples how to generate higher-level context information from low-level context by using rules-based method are shown in Table 3.1.

From the table, some higher-level contexts can also be called situations. They are generated from lower level context information using rule-based reasoning. From those examples, the rules may be adapted. However, there can be more than one rule to represent the same situation. The question is raised which rule is the most suitable for that situation. Also, determining the numbers of rules for each particular application is challenging.

3.3.4 Fuzzy Logic

The fuzzy logic allows the uncertainty of truth reasoning instead of crisp reasoning (Negnevitsky 2005; Shang and Hossen 2013). It is more convenient for incorporating different human opinions, which can easily describe in linguistic terms, and more adapted to insufficient and ambiguous data. The advantages of fuzzy logic are the usage of language variables and the ability to deal with vague systems. It is not necessary to have an accurate quantitative model to determine appropriate action causing the faster and the simpler program development. Although fuzzy logic might not require an understanding of process but more knowledge can help formulating the rules. Complicated systems may require several iterations to find a set of rules resulting in a stable system. This algorithm is used in artificial intelligence oriented areas especially in a decision support system for e-commerce, life insurance, control system, healthcare, etc.

Fuzzy logic consists of four primary processes including fuzzification, fuzzy rule evaluation, aggregation, and defuzzification. Firstly, all crisp factors as the input variables and output variables are gathered. Secondly, these factors are transferred to fuzzy sets with the membership functions using linguistic variables and values. This process is called fuzzification process. Next, the evaluation of fuzzy rule process is operated by a fuzzy inference process. It is designed based on a set of formulated rules. Then, the results of the consequent membership functions of evaluated fuzzy rule-related sets are combined altogether which is an aggregation process. Finally, the result is presented by converting the membership functions, which is the result of the aggregation of rule consequents. The numerical values are obtained as the crisp output in the defuzzification process.

Fuzzy logic is similar to probabilistic reasoning. Instead of representing the probability, the confidence values represent degrees of membership (Román et al. 2002). For fuzzy logic, the partial truth values are used for describing the real-world scenarios. They are accepted as the naturally imprecise notions that are trustworthy and confident to be captured such as short, tall, dark, etc. This imprecise notion is critical for context information processing. For many applications, the fuzzy reasoning cannot be used as a standalone reasoning technique. It usually requires other techniques as the complement such as rules based, probabilistic or ontological based reasoning. For context-aware applications, several examples employ fuzzy logic to handle uncertainty situations (Ranganathan and Campbell 2003) and to represent context information (Mäntyjärvi and Seppänen 2002).

3.3.5 *Ontology Based Reasoning Method*

As mentioned before, the ontology-based method is based on description logic, which is a family of logic-based knowledge representations. The ontological reasoning is normally supported by representations of semantic web languages such as RDF and OWL. While the advantage of ontological reasoning is that it can be integrated well with ontology modeling, the disadvantage is that the ontological reasoning is not good in finding missing values or ambiguous information comparing to other statistical reasoning techniques. For context-aware applications, the ontological reasoning is used in many applications such as activity recognition (Riboni and Bettini 2009), hybrid reasoning (Lane et al. 2010), event detection (Teymourian et al. 2009), etc.

3.3.6 *Probabilistic Logic*

The probabilistic logic uses the probabilities of the facts related to the problem of making the decisions and understanding of the occurrence of events. For context-aware applications, it is used to combine sensor data from two different sources and solve the conflicts between them. Also, it has been used in access control policies (Román et al. 2002). Dempster-Shafer is commonly used in sensor data fusion for activity recognition because it can connect different pieces of the evidence. Hidden Markov Model (Eddy 1996) is one of the favorite probabilistic techniques that use the observable evidence without directly reading the state to represent the state. It is commonly used for activity recognition of context-aware applications such as situation recognition in a smart home (Brdiczka et al. 2009).

3.4 Context Distribution

Distribution information or services to the users or target entities are crucial to application performance, especially for the large-scale network system. Context information which is inferred from many available sources needs to be delivered with efficient and effective ways in the timely fashion. Two methods are commonly used for context distribution (Perera et al. 2014). Firstly, query method which the context consumers make a query request, so the context management system can use that query to produce results. On the other hand, another method is called subscription method. For this method, the context consumer subscribes to the context management system by describing its requirements. The system will return the results when an event occurs or periodically. More specifically, the consumers can subscribe for a particular sensor to an event. This method is typically used for real-time processing.

3.5 Context Adaptation

As mentioned before, the adaptation method is required for any context-aware application to provide the appropriate response to the users personally and the environment rationally. More specifically, adaptive systems refer to the process of enabling the system to fit its behavior and functionalities to the specific needs. The reason of having context adaptation is that both the contexts and the needs change dynamically. Since the applications or the systems have to adjust their behaviors to those changes which are the principle of context-aware applications, there is the need to understand how the adaptation can be performed. Moreover, the applications or the systems act accordingly to the situations of the user and the environment. Consequently, the definition and identification of situation will be briefly discussed in this section.

3.5.1 Situation Identification

The situation means the higher-level context. It can be considered as the abstract state of affairs which is attractive to the applications (Costa et al. 2006). More specifically, a situation is an abstraction of the events in the real world. It is derived from context and pre-defined hypotheses by the designers and the applications (Ye et al. 2012). The situations are required to provide a straightforward and understandable representation of sensor data to the applications. For large-scale applications, there may be more than thousands of situations need to be recognized. As a result, the applications have principal responsibility to define and manage these situations efficiently. It is also crucial to the applications how different situations are related to each other. The applications should have enough knowledge to know which situations can or cannot occur at the same time. Otherwise, the inappropriate adaptive behavior may occur. Five primary relationships exist between situations including generalization, composition, dependence, contradiction, and temporal sequences (Ye et al. 2012). Generalization means that one situation can be the subset of one situation. For example, "swimming" can be a subset of "doing sport." Therefore "doing sport" can be considered as more general than "swimming." Composition means that one situation can combine with other situations. For example, "watching TV" can be the composition of "in the living room" and "TV is on." Dependence means that the occurrence of one situation is determined by the occurrence of one situation. For example, the situations "it rains outside" and "room is cold" can be considered that after the situation "it rains outside" occurs then the situation "room is cold" will occur. Contradiction means that two situations cannot occur at the same time. For example, the situation "taking a shower" and the situation "having breakfast" cannot occur at the same time for the same person. Finally, temporal sequence means that one situation may occur before, or after another situation. For example, situation "pouring coffee" may take place before or after situation "pouring milk."

It can be seen that the situations highly depend on many factors such as the sensor data, the domain knowledge on environments and individual users, and the applications. Some important concerns for situation identification involve three primary processes including representation, specification, and reasoning. For representation, it is challenging to have logical primitives that can represent the imperfect context information from different types of sensors. Moreover, the logical primitives should be flexible enough to capture incomplete data from faulty sensors for avoiding ambiguous meanings. It is challenging to determine the relevant contexts to a situation and define their different contribution to the different situation. The logical specification of the situation usually can be obtained by the experts or learned from training data. For reasoning, it is challenging to infer situations and their relationships from large numbers of imperfect context information. Adaptation of situations means that the applications or the systems need to have functions or services that can be fine-tuned for different situations to serve the right functions or services to the right users and environment with the right way and at the right time. This adaptation requires a situation model which can promote the evolution of situations because the requirements of the users, environments, and applications can be changed over time.

3.5.1.1 Situation Identification Method

The higher level context information is generally required by applications. Several context models and reasoning techniques are needed as discussed from the previous section. This section introduces techniques used for situation identification so that the identified situation can cope with any evolution of the users, the environment and the application itself. For the early age of situation identification, several specification based methods are used by representing expert knowledge in the form of several logic rules. By applying reasoning techniques, the proper situations from low-level context information from sensors can be inferred. These specification based methods are suitable in the case that there are not many sensors involved and the relationships to identify situation are easy to perform. The traditional specification methods are thus any logic based methods (Loke 2010; Weiser 2004). Then, ontology-based methods are used later on due to its representation capacity (Gu et al. 2004; Chen et al. 2003; Ranganathan et al. 2003). Then, the probability-based methods are used together with logical and ontology-based methods to deal with the uncertainty of sensed context information (Haghighi et al. 2008). After the advances in sensor technologies have brought affordable sensors, it is difficult for specification-based methods to use only expert knowledge to deal with proper specifications of the situations from noisy sensor data. In this case, there is the necessary to have learning based method for identifying the situation. The methods from machine learning and data mining are used to solve this problem as shown in several works of activity recognition in smart environments (Chen et al. 2010; Nazerfard et al. 2010; Sánchez et al. 2008; Wu et al. 2010). Many works aim to propose the methods for identifying and explaining more flexible relationships. For example, the Bayesian derivative models are used for describing dependence relationships (Patterson et al. 2003; Tapia et al. 2004; van Kasteren and Krose 2007). The Markov Models (Hasan et al. 2008; Li et al. 2013; Kawanaka et al. 2006) are normally used for explaining the temporal relationships. The Conditional Random Fields are used for more flexible situation modeling (Vail et al. 2007; Liao et al. 2007), etc. Many common machine learning methods are also widely used for classifying sensor data into situations. The example methods are decision trees (Logan et al. 2007; Bao and Intille 2004; Hwang et al. 2010), neural networks (Yang et al. 2008; Choi et al. 2005), and support vector machines (Adomavicius and Tuzhilin 2011), etc. Learning-based methods require a significant amount of training data to set up a model and estimate their model parameters for achieving good results in situation identification.

3.5.2 Awareness Mechanism

As mentioned before, the context-aware application must have the ability to sense the context and to adapt its behavior to provide the appropriate response to the users and/or the environment. Two important issues are concerned for designing and developing context-aware applications. The first issue is the provisioning of context information which has been discussed before in this chapter. This section only introduces the second issue which is about the methodology to build applications to be able to adapt their behaviors depending on the provided context information with or without the explicit intervention from the users. However, the detail of implementation is out of the scope of this book. The adaptation of the service to the preferences, the expectation, and the needs of each user is one of the most desired features of the context-aware applications. These services are described as the contextual adaptation. A model-driven methodology (Sheng and Benatallah 2005) is widely used to perform awareness. It typically consists of four phases including modeling, composition, transformation, and adaptation. This section introduces only adaptation mechanism used in the existing context-aware applications. The literature shows that three main stereotyped relationships can be considered as the primary mechanisms to achieve adaptation ability (Boudaa et al. 2016; Grassi and Sindico 2007; Kapitsaki et al. 2009; Sheng et al. 2010). Firstly, the binding component. It is an association of binding between context elements and application elements which are capable of being awareness. It allows the information retrieval for the users based on available context information. Secondly, the adaption component. It is a mechanism enabling to change the application behavior by selecting the appropriate behavior from among several behaviors accordingly to the current contextual situation. Finally, the trigger component. It normally consists of two main parts including a set of contextual constraints and a set of actions. The action part will be performed when all constraints are satisfied while a set of actions contains all relevant actions required to response to the users or the environment.

The model-driven methodology has also been used for modeling of the interaction between context information and the service. For example, the extension of existing UML syntax (Sheng and Benatallah 2005) is used by introducing appropriate artifacts to enable the construction of context-aware service models. The derived models consist of class diagrams. The classes mainly correspond to the context and the constructed service. Recently, there is the framework supporting the design of context-aware multichannel web applications (Ceri et al. 2007). This framework not only aims to promote the adaptability for the context-aware applications with the least intervention from the user but also to promote the ease development of applications. The design of application front-end is better to be separated from each other. Context information is added to the application data in the form of metadata, whereas context-aware capabilities are added in the hypertext form. Most of on-going research works in awareness mechanism focus on the generic framework for context-aware applications that can achieve the adaptability appropriately to any application as well as the ease of development.

References

Adomavicius, G., & Tuzhilin, A. (2011). Context-aware recommender systems. In *Recommender systems handbook* (pp. 217–253). New York, USA: Springer.

Alegre, U., Augusto, J. C., & Clark, T. (2016). Engineering context-aware systems and applications: A survey. *Journal of Systems and Software, 117*, 55–83.

Balavalad, K. B., Manvi, S. S., & Sutagundar, A. V. (2009, October). Context aware computing in wireless sensor networks. In *International Conference on Advances in Recent Technologies in Communication and Computing, 2009. ARTCom'09* (pp. 514–516). IEEE.

Baldauf, M., Dustdar, S., & Rosenberg, F. (2007). A survey on context-aware systems. *International Journal of Ad Hoc and Ubiquitous Computing, 2*(4), 263–277.

Bao, L., & Intille, S. S. (2004, April). Activity recognition from user-annotated acceleration data. In *International Conference on Pervasive Computing* (pp. 1–17). Heidelberg: Springer.

Barbero, C., Zovo, P. D., & Gobbi, B. (2011, June). A flexible context aware reasoning approach for iot applications. In *12th IEEE International Conference on Mobile Data Management (MDM), 2011* (Vol. 1, pp. 266–275). IEEE.

Bechhofer, S. (2009). OWL: Web ontology language. In *Encyclopedia of database systems* (pp. 2008–2009). New York, USA: Springer.

Bettini, C., Brdiczka, O., Henricksen, K., Indulska, J., Nicklas, D., Ranganathan, A., et al. (2010). A survey of context modelling and reasoning techniques. *Pervasive and Mobile Computing, 6*(2), 161–180.

Bikakis, A., Patkos, T., Antoniou, G., & Plexousakis, D. (2007). A survey of semantics-based approaches for context reasoning in ambient intelligence. In *Constructing ambient intelligence* (pp. 14–23). Heidelberg: Springer.

Boudaa, B., Hammoudi, S., Mebarki, L. A., Bouguessa, A., & Chikh, M. A. (2016). An aspect-oriented model-driven approach for building adaptable context-aware service-based applications. *Science of Computer Programming.*

Brdiczka, O., Crowley, J. L., & Reignier, P. (2009). Learning situation models in a smart home. *Systems, Man, and Cybernetics, Part B: Cybernetics, IEEE Transactions on, 39*(1), 56–63.

Castelli, G., Mamei, M., Rosi, A., & Zambonelli, F. (2009). Extracting high-level information from location data: the w4 diary example. *Mobile Networks and Applications, 14*(1), 107–119.

Ceri, S., Daniel, F., Matera, M., & Facca, F. M. (2007). Model-driven development of context-aware Web applications. *ACM Transactions on Internet Technology (TOIT), 7*(1), 2.

Chen, C., Das, B., & Cook, D. J. (2010, July). A data mining framework for activity recognition in smart environments. In *2010 Sixth International Conference on Intelligent Environments (IE)* (pp. 80–83). IEEE.

Chen, H., Finin, T., & Joshi, A. (2003). An ontology for context-aware pervasive computing environments. *The Knowledge Engineering Review, 18*(03), 197–207.

Chen, H., Finin, T., Joshi, A., Kagal, L., Perich, F., & Chakraborty, D. (2004). Intelligent agents meet the semantic web in smart spaces. *IEEE Internet Computing, 8*(6), 69–79.

Chen, G., & Kotz, D. (2000). *A survey of context-aware mobile computing research* (Vol. 1, No. 2.1, pp. 2-1). Technical Report TR2000-381, Dept. of Computer Science, Dartmouth College.

Choi, J., Shin, D., & Shin, D. (2005). Research and implementation of the context-aware middleware for controlling home appliances. *IEEE Transactions on Consumer Electronics, 51*(1), 301–306.

Costa, P. D., Guizzardi, G., Almeida, J. P. A., Pires, L. F., & Van Sinderen, M. (2006, October). Situations in conceptual modeling of context. In *2006 10th IEEE International Enterprise Distributed Object Computing Conference Workshops (EDOCW'06)* (p. 6). IEEE.

Doukas, C., Maglogiannis, I., Tragas, P., Liapis, D., & Yovanof, G. (2007). Patient fall detection using support vector machines. In *Artificial intelligence and innovations 2007: From theory to applications* (pp. 147–156). New York, USA: Springer.

Eddy, S. R. (1996). Hidden markov models. *Current Opinion in Structural Biology, 6*(3), 361–365.

Elnahrawy, E., & Nath, B. (2004). Context-aware sensors. In *Wireless sensor networks* (pp. 77–93). Heidelberg: Springer.

Favela, J., Tentori, M., Castro, L. A., Gonzalez, V. M., Moran, E. B., & Martínez-García, A. I. (2007). Activity recognition for context-aware hospital applications: Issues and opportunities for the deployment of pervasive networks. *Mobile Networks and Applications, 12*(2–3), 155–171.

Fogarty, J., Hudson, S. E., Atkeson, C. G., Avrahami, D., Forlizzi, J., Kiesler, S., et al. (2005). Predicting human interruptibility with sensors. *ACM Transactions on Computer-Human Interaction (TOCHI), 12*(1), 119–146.

Gellersen, H. W., Schmidt, A., & Beigl, M. (2002). Multi-sensor context-awareness in mobile devices and smart artifacts. *Mobile Networks and Applications, 7*(5), 341–351.

Grassi, V., & Sindico, A. (2007, September). Towards model driven design of service-based context-aware applications. In *International Workshop on Engineering of Software Services for Pervasive Environments: In Conjunction with the 6th ESEC/FSE Joint Meeting* (pp. 69–74). ACM.

Gu, T., Wang, X. H., Pung, H. K., & Zhang, D. Q. (2004, January). An ontology-based context model in intelligent environments. In *Proceedings of Communication Networks and Distributed Systems Modeling and Simulation Conference* (Vol. 2004, pp. 270–275).

Guan, D., Yuan, W., Lee, S., & Lee, Y. K. (2007, October). Context selection and reasoning in ubiquitous computing. In *The 2007 International Conference on Intelligent Pervasive Computing, 2007. IPC* (pp. 184–187). IEEE.

Hagan, M. T., Demuth, H. B., Beale, M. H., & De Jesús, O. (1996). *Neural network design* (Vol. 20). Boston: PWS publishing company.

Haghighi, P. D., Krishnaswamy, S., Zaslavsky, A., & Gaber, M. M. (2008, October). Reasoning about context in uncertain pervasive computing environments. In *European Conference on Smart Sensing and Context* (pp. 112–125). Heidelberg: Springer.

Halpin, T. (2001). Object role modeling: An overview. White paper, (online at www.orm.net). *gadamowmebulia, 20,* 2007.

Hasan, M. K., Rubaiyeat, H. A., Lee, Y. K., & Lee, S. (2008, February). A reconfigurable HMM for activity recognition. In *ICACT* (Vol. 8, pp. 843–846).

Haussler, D. (1990). *Probably approximately correct learning*. Santa Cruz: University of California, Computer Research Laboratory.

Hightower, J., & Borriello, G. (2001). Location systems for ubiquitous computing. *Computer, 8,* 57–66.

Ho, J., & Intille, S. S. (2005, April). Using context-aware computing to reduce the perceived burden of interruptions from mobile devices. In *Proceedings of the SIGCHI Conference on Human Factors in Computing Systems* (pp. 909–918). ACM.

Hofer, T., Schwinger, W., Pichler, M., Leonhartsberger, G., Altmann, J., & Retschitzegger, W. (2003, January). Context-awareness on mobile devices-the hydrogen approach. In *Proceedings of the 36th Annual Hawaii International Conference on System Sciences, 2003* (pp. 10-pp). IEEE.

Huang, S. H., Wu, T. T., Chu, H. C., & Hwang, G. J. (2008, March). A decision tree approach to conducting dynamic assessment in a context-aware ubiquitous learning environment. In *Fifth IEEE International Conference on Wireless, Mobile, and Ubiquitous Technology in Education, 2008. WMUTE 2008* (pp. 89–94). IEEE.

Hwang, G. J., Chu, H. C., Shih, J. L., Huang, S. H., & Tsai, C. C. (2010). A decision-tree-oriented guidance mechanism for conducting nature science observation activities in a context-aware ubiquitous learning environment. *Educational Technology & Society, 13*(2), 53–64.

Hynes, G., Reynolds, V., & Hauswirth, M. (2009, September). A context lifecycle for web-based context management services. In *European Conference on Smart Sensing and Context* (pp. 51–65). Heidelberg: Springer.

Indulska, J., & Sutton, P. (2003, January). Location management in pervasive systems. In *Proceedings of the Australasian Information Security Workshop Conference on ACSW Frontiers 2003* (Vol. 21, pp. 143–151). Australian Computer Society, Inc..

Kapitsaki, G. M., Kateros, D. A., Prezerakos, G. N., & Venieris, I. S. (2009). Model-driven development of composite context-aware web applications. *Information and Software Technology, 51*(8), 1244–1260.

Kawanaka, D., Okatani, T., & Deguchi, K. (2006). HHMM based recognition of human activity. *IEICE Transactions on Information and Systems, 89*(7), 2180–2185.

Keßler, C., Raubal, M., & Wosniok, C. (2009). Semantic rules for context-aware geographical information retrieval. In *Smart Sensing and Context* (pp. 77–92). Heidelberg: Springer.

Ko, K. E., & Sim, K. B. (2008). Context aware system based on Bayesian network driven context reasoning and ontology context modeling. *International Journal of Fuzzy Logic and Intelligent Systems, 8*(4), 254–259.

Kohonen, T., Schroeder, M. R., & Huang, T. S. (2001). *Self-organizing maps*. New York: Springer.

Korel, B. T., & Koo, S. G. (2010). A survey on context-aware sensing for body sensor networks. *Wireless Sensor Network, 2*(08), 571.

Lahiri, S. (2005). *RFID sourcebook*. IBM press.

Lane, N. D., Miluzzo, E., Lu, H., Peebles, D., Choudhury, T., & Campbell, A. T. (2010). A survey of mobile phone sensing. *Communications Magazine, IEEE, 48*(9), 140–150.

Larose, D. T. (2005). k-nearest neighbor algorithm. In *Discovering knowledge in data: An introduction to data mining* (pp 90–106).

Li, H., Yi, Y., Li, X., & Guo, Z. (2013). Human activity recognition based on HMM by improved PSO and event probability sequence. *Journal of Systems Engineering and Electronics, 24*(3), 545–554.

Liao, L., Fox, D., & Kautz, H. (2007). Extracting places and activities from gps traces using hierarchical conditional random fields. *The International Journal of Robotics Research, 26*(1), 119–134.

Lin, T. N., & Lin, P. C. (2005, June). Performance comparison of indoor positioning techniques based on location fingerprinting in wireless networks. In *International Conference on Wireless Networks, Communications and Mobile Computing, 2005* (Vol. 2, pp. 1569–1574). IEEE.

Llinas, J., & Hall, D. L. (1998, May). An introduction to multi-sensor data fusion. In *Proceedings of the 1998 IEEE International Symposium on Circuits and Systems, 1998. ISCAS'98* (Vol. 6, pp. 537–540). IEEE.

Logan, B., Healey, J., Philipose, M., Tapia, E. M., & Intille, S. (2007, September). A long-term evaluation of sensing modalities for activity recognition. In *International conference on Ubiquitous computing* (pp. 483–500). Heidelberg: Springer.

Loke, S. W. (2010). Incremental awareness and compositionality: A design philosophy for context-aware pervasive systems. *Pervasive and Mobile Computing, 6*(2), 239–253.

Mäntyjärvi, J., & Seppänen, T. (2002). Adapting applications in mobile terminals using fuzzy context information. In *Human computer interaction with mobile devices* (pp. 95–107). Heidelberg: Springer.

Mathie, M. J., Celler, B. G., Lovell, N. H., & Coster, A. C. F. (2004). Classification of basic daily movements using a triaxial accelerometer. *Medical & Biological Engineering & Computing, 42*(5), 679–687.

Mcheick, H. (2014). Modeling context aware features for pervasive computing. *Procedia Computer Science, 37,* 135–142.

Nazerfard, E., Das, B., Holder, L. B., & Cook, D. J. (2010, November). Conditional random fields for activity recognition in smart environments. In *Proceedings of the 1st ACM International Health Informatics Symposium* (pp. 282–286). ACM.

Negnevitsky, M. (2005). *Artificial intelligence: A guide to intelligent systems*. Pearson Education.

Nilsson, M., Hjelm, J., & Ohto, H. (2000). Composite capabilities/preference profiles: Requirements and architecture. *W3C Working Draft, 21,* 2–28.

Noy, N. F., & McGuinness, D. L. (2001). Ontology development 101: A guide to creating your first ontology.

Nurmi, P., & Floréen, P. (2004). Reasoning in context-aware systems. Helsinki Institute for Information Technology, Position Paper.

Olifer, N., & Olifer, V. (2005). *Computer networks: Principles, technologies and protocols for network design*. Wiley Publishing.

Pan, J. Z. (2009). Resource description framework. In *Handbook on ontologies* (pp. 71–90). Heidelberg: Springer.

Park, H. S., Oh, K., & Cho, S. B. (2011). Bayesian network-based high-level context recognition for mobile context sharing in cyber-physical system. *International Journal of Distributed Sensor Networks*.

Patel, S. N., Robertson, T., Kientz, J. A., Reynolds, M. S., & Abowd, G. D. (2007, September). At the flick of a switch: Detecting and classifying unique electrical events on the residential power line (nominated for the best paper award). In *International Conference on Ubiquitous Computing* (pp. 271–288). Heidelberg: Springer.

Patterson, D. J., Liao, L., Fox, D., & Kautz, H. (2003, October). Inferring high-level behavior from low-level sensors. In *International Conference on Ubiquitous Computing* (pp. 73–89). Heidelberg: Springer.

Perera, C., Zaslavsky, A., Christen, P., & Georgakopoulos, D. (2014). Context aware computing for the internet of things: A survey. *Communications Surveys & Tutorials, IEEE, 16*(1), 414–454.

Pietschmann, S., Mitschick, A., Winkler, R., & Meißner, K. (2008, December). Croco: Ontology-based, cross-application context management. In *Third International Workshop on Semantic Media Adaptation and Personalization, 2008. SMAP'08* (pp. 88–93). IEEE.

Ranganathan, A., & Campbell, R. H. (2003, June). A middleware for context-aware agents in ubiquitous computing environments. In *Middleware 2003* (pp. 143–161). Heidelberg: Springer.

Ranganathan, A., McGrath, R. E., Campbell, R. H., & Mickunas, M. D. (2003). Use of ontologies in a pervasive computing environment. *The Knowledge Engineering Review, 18*(03), 209–220.

Reignier, P., Brdiczka, O., & Crowley, J. L. (2009). Learning situation models in a smart home. *IEEE Transactions on Systems, Man, and Cybernetics, Part B: Cybernetics, 39*(1).

Riboni, D., & Bettini, C. (2009). Context-aware activity recognition through a combination of ontological and statistical reasoning. In *Ubiquitous Intelligence and Computing* (pp. 39–53). Heidelberg: Springer.

Román, M., Hess, C., Cerqueira, R., Ranganathan, A., Campbell, R. H., & Nahrstedt, K. (2002). A middleware infrastructure for active spaces. *IEEE Pervasive Computing, 1*(4), 74–83.

Rumbaugh, J., Jacobson, I., & Booch, G. (2004). *The unified modeling language reference manual*. Pearson Higher Education.

Sánchez, D., Tentori, M., & Favela, J. (2008). Activity recognition for the smart hospital. *IEEE Intelligent Systems, 23*(2), 50–57.

Schilit, B., Adams, N., & Want, R. (1994, December). Context-aware computing applications. In *First Workshop on Mobile Computing Systems and Applications, 1994. WMCSA 1994* (pp. 85–90). IEEE.

Schmidt, A., Strohbach, M., Van Laerhoven, K., Friday, A., & Gellersen, H. W. (2002, September). Context acquisition based on load sensing. In *International Conference on Ubiquitous Computing* (pp. 333–350). Heidelberg: Springer.

Shang, K., & Hossen, Z. (2013). *Applying fuzzy logic to risk assessment and decision-making*. Canadian Institute of Actuaries.

Sheng, Q. Z., & Benatallah, B. (2005, July). ContextUML: A UML-based modeling language for model-driven development of context-aware web services. In *International Conference on Mobile Business (ICMB'05)* (pp. 206–212). IEEE.

Sheng, Q. Z., Yu, J., Segev, A., & Liao, K. (2010). Techniques on developing context-aware web services. *International Journal of Web Information Systems, 6*(3), 185–202.

Shtykh, R. Y., & Jin, Q. (2008, October). Capturing user contexts: Dynamic profiling for information seeking tasks. In *3rd International Conference on Systems and Networks Communications, 2008. ICSNC'08* (pp. 365–370). IEEE.

Shulsky, A. N., & Schmitt, G. J. (2002). *Silent warfare: Understanding the world of intelligence*. Potomac Books, Inc.

Strang, T., & Linnhoff-Popien, C. (2004, September). A context modeling survey. In *Workshop Proceedings*.

Tapia, E. M., Intille, S. S., & Larson, K. (2004, April). Activity recognition in the home using simple and ubiquitous sensors. In *International Conference on Pervasive Computing* (pp. 158–175). Heidelberg: Springer.

Teymourian, K., Streibel, O., Paschke, A., Alnemr, R., & Meinel, C. (2009, December). Towards semantic event-driven systems. In *2009 3rd International Conference on New Technologies, Mobility and Security (NTMS)* (pp. 1–6). IEEE.

Vail, D. L., Veloso, M. M., & Lafferty, J. D. (2007, May). Conditional random fields for activity recognition. In *Proceedings of the 6th International Joint Conference on Autonomous Agents and Multiagent Systems* (p. 235). ACM.

van Kasteren, T., & Krose, B. (2007, September). Bayesian activity recognition in residence for elders. In *3rd IET International Conference on Intelligent Environments, 2007. IE 07* (pp. 209–212). IET.

Van Laerhoven, K. (2001). Combining the self-organizing map and k-means clustering for on-lineclassification of sensor data. *Artificial Neural Networks—ICANN 2001* (pp. 464–469).

Wang, X. H., Zhang, D. Q., Gu, T., & Pung, H. K. (2004, March). Ontology based context modeling and reasoning using OWL. In *Proceedings of the Second IEEE Annual Conference on Pervasive Computing and Communications Workshops, 2004* (pp. 18–22). IEEE.

Weiser, M. (2004). Reasoning about uncertain contexts in pervasive computing environments.

Wu, C., Khalili, A. H., & Aghajan, H. (2010, August). Multiview activity recognition in smart homes with spatio-temporal features. In *Proceedings of the Fourth ACM/IEEE International Conference on Distributed Smart Cameras* (pp. 142–149). ACM.

Yamabe, T., Takagi, A., & Nakajima, T. (2005, August). Citron: A context information acquisition framework for personal devices. In *11th IEEE International Conference on Embedded and Real-Time Computing Systems and Applications (RTCSA'05)* (pp. 489–495). IEEE.

Yang, J. Y., Wang, J. S., & Chen, Y. P. (2008). Using acceleration measurements for activity recognition: An effective learning algorithm for constructing neural classifiers. *Pattern Recognition Letters, 29*(16), 2213–2220.

Yanwei, S., Guangzhou, Z., & Haitao, P. (2011, December). Research on the context model of intelligent interaction system in the internet of things. In *2011 International Symposium on IT in Medicine and Education (ITME)* (Vol. 2, pp. 379–382). IEEE.

Ye, J., Dobson, S., & McKeever, S. (2012). Situation identification techniques in pervasive computing: A review.*Pervasive and mobile computing, 8*(1), 36–66.

Zhao, F., & Guibas, L. J. (2004). *Wireless sensor networks: An information processing approach.* Morgan Kaufmann.

Chapter 4
Communications for Context-Aware Applications

Abstract This chapter accounts for the foundation of communication for context-aware applications. Some relevant communications are introduced including communication network, sensor network, body area network, and social network. For communication networks, the foundation of the communication system is described together with the current popular wireless systems. The sensor network is introduced to provide the whole idea of how to manage many different types of sensors. Body area network becomes popular nowadays, especially for healthcare application domain. Since this network involves a lot of small size of body sensors, some significant concerns for gathering body context is discussed in this chapter. Finally, the social network is introduced as the important communication for people nowadays. Understanding the social behavior appropriately enables to provide the appropriate feedback to the group and the individual user. Some measurements of social network analysis are also shown in this chapter.

Context-aware applications rely on the communication system which any entities can be connected such as the applications connect to the user, the users connect to the devices, the users connect to the users and the devices connect to the devices, etc. This chapter aims to introduce the foundation for some important communication networks being involved in many context-aware applications. They are classified based on the interactions among entities of the context-aware applications. Therefore four types of communication networks are discussed in this chapter. Wireless Communication Network is firstly introduced because it is the common communication system among the users, the hardware, and the applications. It involves the way of transmitting context information or any response to other relating components. Secondly, Sensor Network is introduced. It is responsible for the communications among sensors to gather real world information. It involves the context acquisition process. Next, the Body Area Network, which is in charge of the communications among body sensors to collect physical information of the users, is introduced. Finally, the Social Network is presented. It represents not only the interaction among the users but also among the entities. The details of all networks and the significant concerns including the existing applications will be

P. Temdee and R. Prasad, *Context-Aware Communication and Computing: Applications for Smart Environment*, Springer Series in Wireless Technology,
DOI 10.1007/978-3-319-59035-6_4

shown and discusses in this chapter. At the end of this chapter, the reader will have a clearer concept of what types of communication networks should be involved in their context-aware applications.

4.1 Communication Networks

Most of the context-aware applications rely heavily on wireless communication. Therefore, this section aims to introduce wireless communication (Wood et al. 2008). Wireless communication covers many networks such as Wireless Local Area Network (WLAN), Wireless Metropolitan Area Network (WMAN), Bluetooth, Wireless Sensor Network (WSN), Zigbee, RFID, First Generation mobile network (1G), Second-Generation mobile network (2G), Third-Generation mobile network (3G), General Packet Radio Service (GPRS), etc. Wireless communication provides convenience for the users more than wired communication. Wired communication technologies enable users to access a server remotely but have the limitation that the users have to stay at locations that can reach the computer with wired connections. On the other hand, the wireless communication allows the users to get services anywhere under the coverage area. This chapter will introduce the foundation and necessity of wireless communication system promoting the efficiency of context-aware applications.

4.1.1 Communication Systems

Before going to more detail of wireless communication networks. The concept of the communication system, in general, is worth to be introduced. The communication system is a system model describing a communication exchange between two stations including the source and the destination or the transmitter and the receiver. The signals or information passes from the source to the destination through the channel. Before transmitting, the signals must be firstly processed by several stages including signal representation, signal shaping, encoding, and modulation. The signals may face different types of impairment to cross the channels such as noise, attenuation, and distortion. The purpose of a communication system is to carry information from one point to another. A typical communication system consists of three main components including source, channel, and destination as shown in Fig. 4.1. The information sources are any source that can provide information such as audio, image, text, data, etc. The channels are the mediums used for transferring signal from the source to the destination.

More comprehensive view of the transmission system is shown in the conceptual diagram as illustrated in Fig. 4.2. The communication system consists of an input transducer, transmitter, channel, receiver, and output transducer.

Fig. 4.1 Main components of communication system

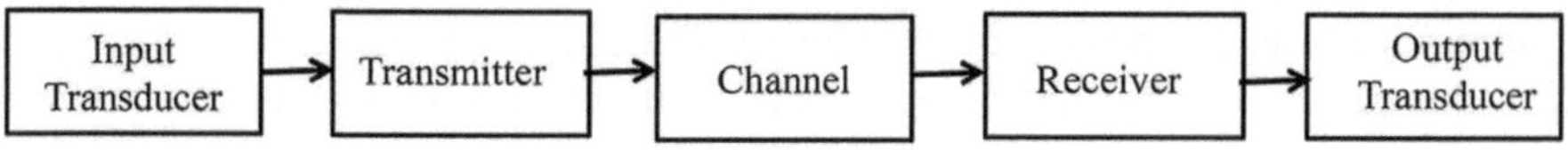

Fig. 4.2 Communication system conceptual diagram

From Fig. 4.2, the input transducer converts source such as microphone, camera, keyboard, etc. to electric signals. The transmitter converts the electrical signal into the form that is suitable for being transmitted through the channel. It includes the modulator and the amplifier for manipulating the signal to have proper shape and magnitude. Channels which are the mediums used to transfer signal from transmitter to the receiver which can be point-to-point or broadcasting and can be wired and wireless channels. Wired communication channels are such as twisted pair, cable, fiber optics, etc. Wireless communication is a transmission of electromagnetic waves from the antenna to the antenna that the propagation characteristics vary with frequency. The receiver estimates the output from the original transducer output. It also includes demodulator and amplifier. The output transducer finally converts the signal into the usable form of information such as the speaker, monitor, etc. Transmitters and receivers are designed to overcome the distortion and noise.

4.1.2 Wireless Communication and Networks

For decades, there has been extensive research works in the field of wireless communication. Wireless communication is considered to be the fastest growing communication industry. More specifically, many networks at home, office or the campus are replaced by wireless local area networks currently. Many new context-aware applications have emerged from research to commercial products such as automated factories, smart homes and appliances, remote telemedicine, etc. The dramatically growth of wireless systems and the portable personal smart devices indicate a future for context-aware applications. This section will briefly review the history of wireless networks from cellular, satellite, and other current wireless networks.

4.1.2.1 History of Wireless Communications

In the pre-industrial age, the wireless networks were firstly developed in the form of the information transmission over line-of-sight distances by using different types of signals such as smoke signals, torch signal, or flashing mirrors. The combination of different types of signal was created to convey more complex messages. Telescopes extend this way of communication. Observation stations were built along the way such as on the hill and along the road to relay these messages over the long distances. Later, the telegraph network invented by Samuel Morse in 1838 was introduced to replace all first communication networks. In 1895, the first radio transmission was demonstrated by Marconi. The radio communications were born with the advanced radio technology enabling transmissions over larger distances with better quality, less power, smaller and cheaper devices. Consequently, the public and private radio communications, the television, and the wireless network were more affordable.

At the early stage, the radio systems transmitted analog signals. Currently, most of them transmit digital signals that can be obtained directly from the digital data signal or by digitizing the analog signal. The digital radio system groups the binary bits into packets called a packet radio. The radio is always idle except when it transmits a packet. ALOHANET, was developed at the University of Hawaii in 1971, is the first network using packet radio. It employed radio transmission to enable computer sites of seven campuses to communicate with a central computer, and it had a star topology with the central computer as a hub (Goldsmith 2005). Later, packet radio networks were widely found in militarily usages and commercial applications for wide-area wireless data services which firstly introduced in the early 1990s. It enables different types of wireless data access with 20 Kbps data rate such as email, file transferring, and web browsing, etc. These services were later disappeared in the late of 1990s because of low data rate but high cost.

In 1970s, wired Ethernet technology has been introduced causing many commercial products away from the radio-based networking. With 10 Mbps data rate through cables, it enabled various kinds of useful applications. Although wired Ethernets can offer data rates of 100 Mbps currently, wireless LANs(WLANs) are more preferred in many homes, offices, and campus environments because of more convenient and freedom over wired networks. So far, the cellular telephone system is considered to be the most successful application of wireless networking. The successful history of cellular networks was begun during the 50s and 60s when the the cellular concept was introduced by AT&T Bell Laboratories (Goldsmith 2005) The important feature of cellular network is the power of a transmitted signal falling off with distance. Two users can operate on the same frequency at different locations spatially with the minimal interference between them. Consequently, a larger number of users can be obtained. In the early 1990s, the second generation of cellular systems based on digital communications was developed with higher capacity. The digital hardware had been improved to support this new generation technology such as cost, speed, and power efficiency. The second generation cellular systems supported mainly voice services. These systems were evolved to support different kinds of data services such as email, Internet access, and short messaging (Goldsmith 2005). Currently, the third generation is still employed although the forth generation is widly implemented. The fifth generation will be launched in the short future.

4.1.3 *Current Wireless Systems*

The examples of wireless systems are cellular networks, WLANs, Satellite Systems, Paging Systems, Broadband Wireless Access, Ultra-wideband Radios, Low-Cost Low-Power Radios including Bluetooth and Zigbee, etc. This section provides a brief overview of some important current wireless systems which are widely found in many context-aware applications.

4.1.3.1 Cellular Telephone Systems

As mentioned before that cellular telephone systems are hugely popular worldwide for decades, these systems also have primary role for wireless revolution. Cellular systems provide two-way data communication between many levels of area coverage including regional, national, or international coverage. Figure 4.3 shows the cellular structure and frequency allocation for each cell. Cells are the hexagonal shape having the base station located in the middle. At the base station, there are transmitter, receiver, and control unit. The radius of the cell is determined by the power of the base station. Specifically, some sets of channels (A-F) are assigned to each cell. The same channel set can be used in another cell which is some distance away, as shown in Fig. 4.3.

The centralized base station controls all operations within the cell. The spatial separation of cells that will be reused by the same channel set can be considered as the most concern. This separation should be as small as possible so that the frequencies can be reused as often as possible without intercell interference. The intercell interference is the interference that the users in different cells operate on the same channel set. It may increase as the reused distance decreases because of the smaller propagation distance between interfering cells. Although the smaller cell sizes can increase the network capacity, it can also increase the interference. The accurate signal propagation within the cells is mainly considered for determining the placement of base station. For early designs, the cell base stations were typically driven by the high cost of base stations which were placed on the tall buildings or

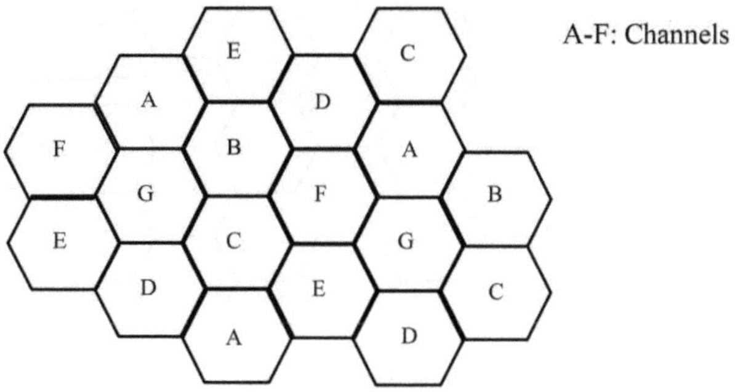

Fig. 4.3 Cellular structure and frequency spectrum allocation

mountains with large cell sizes approximately 6 miles in diameter. These large cells are called macrocells which usually were used in remote areas together with high-power transmitters and receivers.

For the cellular systems in urban areas, the smaller cells with base stations close to street level are commonly used for transmission of lower power. Those little cells are called microcells or picocells. Microcells have small coverage area with approximately a half mile in diameter. Low power transmitters and receivers are used to avoid interference between cells in other clusters. Picocells cover areas such as the building, the tunnel, or the exhibition center, etc. The main reason for the evolution of smaller cells is the need of higher capacity in the areas having more user density. Consequently, the smaller size with the lower cost of base stations is necessary. Moreover, less power is required since the terminals are closer to the base stations. However, the smaller cells still require sophisticated network design.

For given geographical areas, all base stations are connected by a high-speed communications link to a central controller of a network called Mobile Telephone Switching Office (MTSO) (Goldsmith 2005), as shown in Fig. 4.4. The MTSO allocates channels within each cell, coordinates handoffs between cells when the mobiles travel through cell boundaries, and routes the calls to and from the mobile users. The MTSO can route voice calls through the Public Switched Telephone Network (PSTN) or the Internet access (Goldsmith 2005).

For the early stage of cellular systems in 1960, the analog communications are mainly used. Then, the next generation systems move to digital communication because of many advantages. First of all, the components of the networks are cheaper, faster and smaller. They usually require less power. With the advance technology, the digital system can use efficient compression techniques and error correction methods for improving voice quality. In term of capacity, the digital systems have higher capacity than analog systems because of the efficient digital modulation techniques for sharing the cellular spectrum. Also, digital systems can offer addition data services besides voice service such as short messaging, email, Internet access, etc. However, the users can also experience poor voice quality, call dropping, and spotty coverage in some particular areas. As mentioned before, the

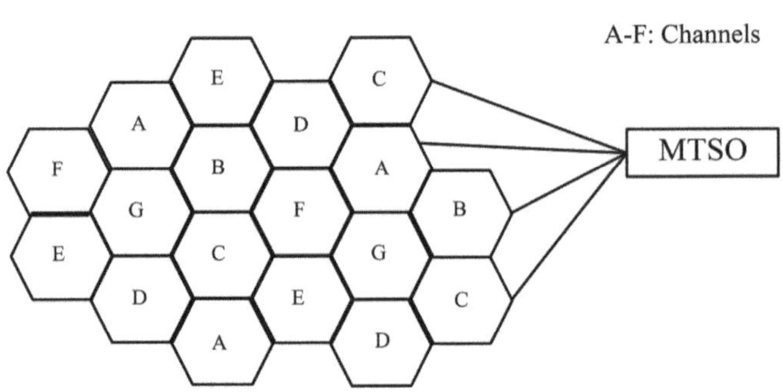

Fig. 4.4 Cellular structure and MTSO

main aim of the cellular system design is to utilize the capacity of the channel particular for handling as many calls as possible for a given bandwidth with some degree of quality of service. As a result, the spectral sharing in communication systems is required. Spectral sharing is also called multiple access. It is used in communication systems and can be done by dividing the signal dimensions into different dimension such as the frequency, time, and code space dimension as frequency-division multiple accesses (FDMA), time-division multiple accesses (TDMA), and code-division multiple accesses (CDMA) respectively.

For frequency-division multiple accesses (FDMA), it is the initial multiple-access technique for the cellular systems. The total bandwidth of the system is divided into many orthogonal frequency channels. While having a call, each user is assigned a pair of frequencies. More specifically, one frequency is used for downlink and onefor uplink. The same allocated frequency pair is not used in the same cell or adjacent cells during the call to avoid the channel interference. However, the FDMA channel carries only one phone circuit at the time. The basic concept of FDMA is shown in Fig. 4.5a. For time-division multiple access (TDMA), the time is divided orthogonally instead of the frequency. Each channel occupies the whole frequency band over its assigned timeslot. It can be seen that the continuous transmission is not required for the digital systems because the users do not always use the allocated bandwidth. TDMA can be considered as a complimentary access technique to FDMA. It is harder to implement TDMA than FDMA since the users must be synchronized in time. However, it is much easier to accommodate multiple data rates with TDMA since multiple time slots can be assigned to a user. The basic concept of TDMA is shown in Fig. 4.5b. For code-division multiple access (CDMA), the same bandwidth is occupied by all users. They are assigned with separate codes distinguishing them from each other. CDMA uses specific random numbers to encode bits of information. The only limitation of the system is the computing process of the base station and its ability to separate noise from the actual data. The basic concept of CDMA is shown in Fig. 4.5c.

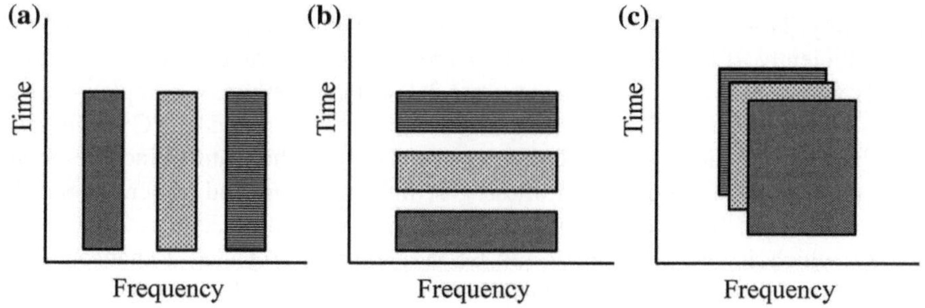

Fig. 4.5 Basic concepts of FDMA, TDMA, and CDMA

Revolution of Cellular Telephone Network

The cellular telephone network is a radio-based technology. Radio waves are electromagnetic waves that antennas propagate. Most signals are in the 850, 900, 1800, and 1900 MHz frequency bands. In the middle of 1940s, car-based telephones were firstly introduced. The single large transmitter was placed on top of a tall building. Single channel was used for sending and receiving. To talk, the users push a button to enable transmission and disable reception. Later in the 1950s, this system was known as "push-to-talk." It was used by taxis and police cars. Later in the 1960s, the Improved Mobile Telephone System (IMTS) was introduced. This technology used two channels for sending and receiving respectively. Therefore, there is no need for push-to-talk anymore. More specifically, it used 23 channels from 150 to 450 MHz. The cellular network in this early stage can be considered as zero generation.

The cellular network can be considered as the fastest growing sector of communication industry since 1982. The first generation, 1G technology, refers to mobile telecommunications which were first introduced in the 1980s. It is voice-oriented systems based on analog technology. It allows the voice calls and uses analog signal. It has the speed up to 2.4 kbps. The well-known examples are Advanced Mobile Phone Systems (AMPS) and cordless systems. The AMPS was first launched in the USA using FDMA with 30 kHz FM-modulated voice channels. AMPS was invented at Bell Labs and firstly installed in 1982. The similar technologies were employed in England which is called as Total Access Communication System (TACS) and in Japan which is called as the first generation mobile cellular system (MCS-L1). For 1G technology, the system has many disadvantages (Arjmandi 2016) such as poor voice quality, poor battery life, large phone size, no security, and limited capacity, etc.

The second generation of cellular networks is called 2G technology. This technology is based on digital transmission having different approaches in US and Europe. The popular 2G wireless technology is known as GSM which was launched in Finland in 1991. The 2G network uses digital signals, and its data speed is up to 64 kbps. It enables services such as text messages, picture messages and multi-media message (MMS). It provides the better quality and capacity than 1G technology. However, 2G requires strong digital signals for the mobile phone to work efficiently. If there is no network coverage in some particular areas, the digital signals would be weak. At the same time, 2G still cannot handle complex data as videos. There is also 2.5G technology which can be described as 2G technology combined with General Packet Radio Service (GPRS). The features include phone calls, sending and receiving email messages, web browsing, and camera phone. Its speed is between 64 and 144 kbps. Another enhancement of 2.5G is 2.75G technology which is the technology for GSM evolution so-called Enhanced Data Services for GSM Evolution (EDGE). However, EDGE works only on GSM networks and has the maximum speed of 384 kbps. EDGE also has adaptive techniques to mitigate the fading effect.

In the 2000s, the third generation is called 3G technology was introduced. The data transmission speed has increased to between 144 kbps and 2 Mbps. The phone is typically called smartphone having the features to accommodate web-based applications, audio, and video files. The essential characteristics of the 3G technology are to provide the faster communication for sending and receiving large email messages, to have high-speed web, to have more security, to perform video conferencing and TV streaming, to have the greater capacity and broadband capacity, etc. However, there have also been some disadvantages of 3G technology. For example, 3G license services and phones are typically expensive. It is so challenging to build the infrastructure for 3G technology to support high bandwidth requirement. Moreover, the cell phone is still large size.

The fourth generation is called 4G technology which was started from the late 2000s. It is capable of providing 100 Mbps to 1 Gbps speed. The important features of 4G are multimedia, anywhere and anytime mobile. It supports global mobility by having integrated wireless solutions. It can provide any service at any time for the user requirements and customizes personal services with high Quality of Service (QOS) and high security. In conclusion, 4G provides greater safety, higher speed, higher capacity and lower cost per bit, etc. However, this technology still requires more battery usages. It is also hard to implement the supporting infrastructure because it needs complicated hardware. The equipment is still expensive.

The fifth generation of the cellular network is called 5G technology started developing since the late 2010s and will be launched soon. It defines complete wireless technology. It is designed to support Wireless World Wide Web (WWWW). 5G technology provides high speed, high capacity, large broadcasting of data in Gbps. It supports large phone memory, dialing speed, clarity in audio and video. Moreover, it is designed to support interactive multimedia, voice, streaming video, the Internet, etc. especially with High Density (HD) quality. In conclusion, 5G is more efficient and more attractive technology than others. 5G is expected to be available in the market in the year 2020 with affordable cost and more reliability than the previous technologies. However, both 4G and 5G are designed to involve the integration of Local Area Network (LAN), Wide Area Network (WAN), and Personal Area Network (PAN). Especially, PAN extremely has the influence of context-aware applications in various application domains nowadays such as business, industry, education, healthcare, smart vehicle, etc. Some characteristics of each cellular network generation are summarized in Table 4.1.

4.1.3.2 Wireless Local Area Network

Local Area Network (LAN) is a communication network interconnecting a variety of data communicating devices within a small geographic area. It broadcasts data at high data transfer rates and very low error rates. Firstly appeared in the 1970s, LANs have become quickly widespread in many commercial and academic environments. The advantages of LANs are the ability to share and support hardware and software resources, the capacity to secure the transferring at high speeds with

Table 4.1 Characteristic summarization of cellular network generation

Generation	1G	2G	3G	4G	5G
Data bandwidth	2 Kbps	14–16 Kpbs	2 Mbps	200 Mbps	>1 Gbps
Core network	PSTN	PSTN	Packet network	Internet	Internet
Technology	Analog cellular	Digital cellular	CDMA/IP Technology	Unified IP and LAN/WAN/WLAN/PAN	WWWW
Multiplexing	FDMA	TDMA/CDMA	CDMA	CDMA	CDMA
Service	Analog voice	Digital voice, SMS	Integrated higher quality audio, video and data	Higher capacity, Complete IP, Multimedia	Dynamic information access, variable devices with smart capability

low error rates, etc. However, there are some disadvantages of LANs. For example, the equipment and support are costly. Some types of hardware may not be able to interoperate. Also, the maintenance cost continues to grow exponentially. LANs are interconnected by one of some basic configurations including bus/tree topology, star-wired bus topology, star-wired ring topology and Wireless LAN or WLAN. Firstly, Bus or tree topology is the original topology. The workstation has a network interface card (NIC) attaching to the bus which is a coaxial cable via a tap. The data can be transferred using either broadband analog signals or baseband digital signals. While baseband digital signals are bidirectional or two-direction transmission, broadband signals are uni-directional or one direction transmission. Buses can be split and joined for creating the trees. For star-wired bus topology, this topology physical looks like a star but operates logically as a bus. The star design is based on the hub for taking the incoming signal and immediately broadcasting it out all connected links. The hubs can be interconnected through different mediums such as twisted pair, coaxial cable, or fiber optic cable to extend the size of the network. All workstations attach to the hub by using unshielded twisted pair. For star-wired ring topology, this topology physically appears as a star but operates logically as a ring. Star-wired ring topology has Multi-station Access Unit (MAU) acting as a hub. This hub broadcasts all incoming signals onto all connected links. Then, the MAU passes the signal around in a ring. To increase the size of the network, MAUs can also be interconnected. For WLANs, there is no a particular topology because a workstation in WLANs can be anywhere within the transmitting distance to an access point. Two essential components for WLANs are the client radio and the access point. The client radio is usually a personal computer (PC) card with an integrated antenna. The access point (AP) is an Ethernet port with a transceiver acting as a bridge between the wired and wireless networks and performing basic routing functions. A connection between the client and the user in WLANs is accomplished by the wireless medium such as an Infrared (IR) and a Radio

Frequency (RF) communications instead of any wired medium. This connection allows the remote users stay connected to the network although their devices are not physically attached to the network. Wireless connections are commonly connected through handheld devices with a built-in RF interface. The important feature of WLANs is that they can be used independently from the wired networks. The network spectrum for communications is designed with the free license in 2.4–2.5 GHz band.

WLANs aim to provide high-speed data for a small region, such as a small building, a campus, etc. The users can move from place to place and still be connected. Wireless devices are typically stationary or moving at pedestrian speeds. WLANs connect local computers approximately 100 m range and the data is broken into packets. Channel access is shared which is called random access. WLANs are flexible for applications requiring the mobility. In 1985, the Federal Communications Commission (FCC) enabled the commercial development of WLANs by authorizing frequency band for WLAN products called the Industrial, Scientific, and Medical (ISM) frequency band (Goldsmith 2005). The WLAN vendors do not need to obtain the license for operating in this band. The original unlicensed bands are the ISM bands at 900 MHz, 2.4 GHz, and 5.8 GHz.

In the early 1990s, the first generation WLANs appeared as incompatible protocols. Most of them operates within the 26 MHz spectrum of the 900 MHz ISM band and uses direct sequence spread spectrum with the data rates of 1–2 Mbps. For architecture, both star and peer-to-peer can be found. Because of lacking standardization, these products have high development costs, low-volume production, and small markets. The second generation of WLANs in USA operates with 80 MHz of spectrum in the 2.4 GHz ISM band (Goldsmith 2005). The WLANs standard for this frequency band is IEEE 802.11b. This standard was developed to avoid some problems of the first generation systems. The standard specifies the direct sequence spread spectrum which has around 1.6 Mbps data rate and an approximately 150 m range. The network architecture can be found either star or peer-to-peer topology, although the star feature is more popular. Many laptops come with integrated 802.11b WLAN cards. Many organizations, shops, and places have installed 802.11b base stations throughout their locations to offer free wireless accessing.

A wireless network includes some essential components including LAN adapter, access points, outdoor LAN Bridge. LAN adapter is made in the same fashion as wired adaptors such as Personal Computer Memory Card International Association (PCMCIA) card bus, Peripheral Component Interconnection (PCI), and Universal Serial Bus (USB). They enable users to access the network. An access point (AP) is the wireless equivalent of a LAN hub. It receives, buffers, and transmits data between the WLANs and the wired network. Outdoor LAN Bridge is used to connect LANs in other buildings. WLANs have many advantages especially the improvement of productivity with real-time access to information regardless of the location of the users. However, there are some issues to concern for deploying WLANs such as frequency allocation, interference, reliability, security, power consumptions, mobility, and throughput.

There have been five major protocols for wireless communication including 802.11, 802.11a, 802.11b, 802.11g, and 802.11n. Firstly, the 802.11 standards were released in 1997, and it is now no longer use. It is considered as the original wireless protocol. It has low interoperability because of loose specifications. The Frequency Hopping Spread Spectrum (FHSS) and Direct-Sequence Spread Spectrum (DSSS) for modulation are used (Kao 2002). Two additional standards in the 802.11 families developed to provide higher data rates are 802.11a and 802.11b. The IEEE 802.11a standard is released in the late 1999 and operates with 300 MHz of spectrum in the 5 GHz band. It uses multicarrier modulation and has the data rates between 20–70 Mbps. It provides bandwidth up to 54 Mbps and uses Orthogonal Frequency-Division Multiplexing (OFDM) to transmit a signal over several sub-signals for higher efficiency. In 1999, the 802.11b standard was introduced. It provides bandwidth up to 11 Mbps and uses Direct-Sequence Spread Spectrum (DSSS) to transmit a signal over several sub-signals for higher efficiency. It operates within the 2.4 GHz band. Since 802.11a has more bandwidth and consequently has many more channels than those of 802.11b, it can support more users at the higher data rates (Goldsmith 2005). For 802.11g, it uses multicarrier modulation in 2.4 GHz with speeds up to 54 Mbps or 108 Mbps with particular implementations. It was released in the middle of 2003. However, it is adopted quickly after releasing of cheap and high bandwidth. Also, it can be considered as the most conventional wireless network at that time. In 2009, 802.11n was just released. It is considered as the newest member of the 802.11 families. It can be used in either 2.4 GHz or 5 GHz bands with up to 600 Mbps bandwidth. It employs OFDM which uses higher frequencies for increasing the number of carrier waves. Additionally, Multiple Input Multiple Output (MIMO) is introduced for supporting higher efficiency. Many current WLAN cards and access points support all standards to avoid incompatibilities. Table 4.2 shows the characteristic summarization of IEEE 802.11 standard family.

4.1.3.3 Wireless Personal Area Network

One of the most popular context-aware applications is about smart environment such as home, office, hospital, airport, etc., various kinds of small sensors are involved. To enable these small sensors to be able to communicate with each other or with other devices in the short range, Wireless personal area networks (WPANs) have been playing the leading role. WPANs are commonly used for conveying information among groups of participant devices over short distances. The need of WPANs has been increased for many reasons especially when the users want to interconnect the portable computers and the devices like peripherals and sensors. These devices may be carried or worn by a person and may be located nearby. Therefore, a WPAN is a short-distance wireless network specifically designed to support portable and mobile computing devices such as PCs, portable devices, wireless printers, storage devices, or any other devices. The history of WPANs started when IEEE 802.15 working group was established in March 1999. Unlike

Table 4.2 Summarization of some standards in IEEE 802.11 WLAN Family

Generation	802.11	802.11a	802.11b	802.11g	802.11n
Released year	1997	Late 1999	1999	Mid of 2003	2009
Bandwidth (Mbps)	<2	54	11	54	600
Frequency (GHz)	2.4	5	2.4	2.4	2.4 or 5
Multiplexing	DSSS, FHSS	OFDM	DSSS	OFDM	MIMO-OFDM

WLANs, a connection made through a WPAN involves little or no infrastructure to the outside world. Consequently, this connection allows small, power-efficient, inexpensive solutions. The widely used WPANs are Bluetooth and Zigbee which are examples of wireless communication with low-cost and low-power radio. These technologies have the common goals in getting rid of cable connections, having little involvement or no infrastructure, and dealing with the interoperability of many devices. However, they are different in some characteristics.

Bluetooth is developed to answer the need for short-range wireless connectivity by supporting ad hoc network and interoperability requirements without the cable. Ad hoc network means that the device with Bluetooth radio can establish the connection with another device when they are in range. Bluetooth is the technology promoted by Ericsson in Sweden and Nokia in Finland. Currently, Bluetooth is a standardized protocol for sending and receiving data via a 2.4 GHz wireless link. Moreover, it is a secure protocol. The standard published by an industry consortium is known as IEEE 802.15.1. It can support data, audio, graphics, and videos. The Bluetooth devices are recognized and speak each other in the same way as a computer does with the printer. The key features of Bluetooth are less complication, less power consumption, low cost, and high robustness.

ZigBee is the technological standard created for sensor networks. It is developed in 1998 when Bluetooth is considered that it may not be suitable for many applications, in particular for sensor networks. It can be seen as simpler as and cheaper than Bluetooth. The primary objectives of ZigBee are ease of installation, reliable data transfer, short-range operation, low cost, and reasonable battery life. Therefore, it is straightforward but flexible protocol. The maximum raw data rate can be 250 kbps. With IEEE 802.15.4 standard, ZigBee has excellent performance in low Signal to Noise Ratio (SNR) environments. In 2002, it is created by ZigBee Alliance which is organized as a nonprofit corporation. The primary responsibility of ZigBee Alliance is to build specification, certify the programs, and develop branding, market and user education. ZigBee operates in unlicensed bands such as ISM 2.4 GHz Global Band at 250 kbps, 868 MHz European Band at 20 kbps, and 915 MHz North American Band at 40 kbps (Hussain et al. 2015). It is widely used for connectivity between small packet devices such as remote control of electrical

Table 4.3 Comparison of Bluetooth and ZigBee characteristics

Technology	Bluetooth	ZigBee
Transmission	Larger packets over small network	Smaller packets over large network
Focused Network	Ad-hoc networks	(Mostly) Static networks
Standard	IEEE 802.15.1	IEEE 802.15.4
Power profile	Days	years
Range	10 m	70–300 m
Data rate	1 Mbps	250 Kbps
Applications	Screen graphics, hands-free audio, mobile phones, headsets, PDAs, etc.	Home automation, toys, remote controls, etc.

appliance, home automation and security, personal healthcare monitoring and diagnosis, etc. Some essential characteristics of Bluetooth and ZigBee are summarized in Table 4.3.

4.2 Sensor Networks

Sensors or transducers convert physical phenomenon such as heat, light, motion, and sound into electrical signals. They are very necessary for the context-aware applications, particularly for context acquisition. A sensor node is a primary unit of the sensor network. This unit contains essential components onboard including sensors, processor, memory, transceiver, and power supply. Recent technology has made the realistic deployment of sensors to be tiny, low-cost, low-power, high capability of local processing and wireless communication. There is the coordination methods required for dealing with a large number of sensor nodes. The sensors are managed properly to measure a given physical environment in complete detail. Therefore, a sensor network is the collection of the sensor nodes coordinating with each other to perform some specific actions. More specifically, a vast number of sensor nodes are deployed either inside or very close to the sensed phenomenon. The sensor networks differ from the traditional networks that they depend on dense deployment and coordination to carry out their tasks. The significant concerns for sensor networks would be the capabilities for distributed processing and low energy communication. The sensor networks previously consist of the small number of sensor nodes that are wired to a central processing station. However, the current sensor network focuses more on wireless and distributed sensing nodes (Estrin et al. 2001) as Wireless Sensor Network (WSNs).

WSNs are the networks consisting of multiple distributed sensors communicating through wireless communication and coordinating together to capture some physical phenomenon. The sensed information is later processed to get the required results. WSNs mainly use broadcast communication while ad hoc networks use point-to-point communication. The capability of WSNs is restricted because of some important issues such as small size, low power, limited memory, and constrained energy. WSNs require protocols and algorithms with self-organizing capabilities to deal with the network reconfiguration. The sensors should be utilized to produce the maximum performance with less power. Additionally, the computation should be done quickly as new data is always generated in the timely fashion.

WSNs involve three leading technologies including embedded, networked and sensing technologies. The embedded technology aims to embed numerous distributed devices to monitor and interact with physical world. The networked technology seeks to coordinate network devices and perform higher-level tasks. Sensing technology aims to exploit both spatially and temporally sensing and provide the appropriate response to the actuator. Figure 4.6 shows the WSNs communication architecture. The main participants in WSNs are data sources, data sinks, and task manager. The data sources are typically equipped with different kinds of actual sensors. The data sinks aim to receive data from the WSNs. They can be part of the WSNs or external entity. Finally, the task manager node seeks to control some devices based on received data.

The applications of WSNs can be found in many different domains, especially in automatic monitoring applications. The popular areas of control applications are in environmental and habitat monitoring, precision agriculture, military surveillance, healthcare monitoring, intelligent alarms, traffic management and surveillance, etc. For example, the information collected from various sensors for smart alarm at home or in the office is used for making the decision whether the alarm signal should be allowed accordingly to the current condition. For precision agriculture, WSNs are employed to help making agricultural operations more efficient, while reducing the environmental impact and investment. The information collected from the sensors can be used to evaluate optimum sowing density, estimate fertilizers, and other necessary inputs, to obtain more accurately crop yields, etc. For the

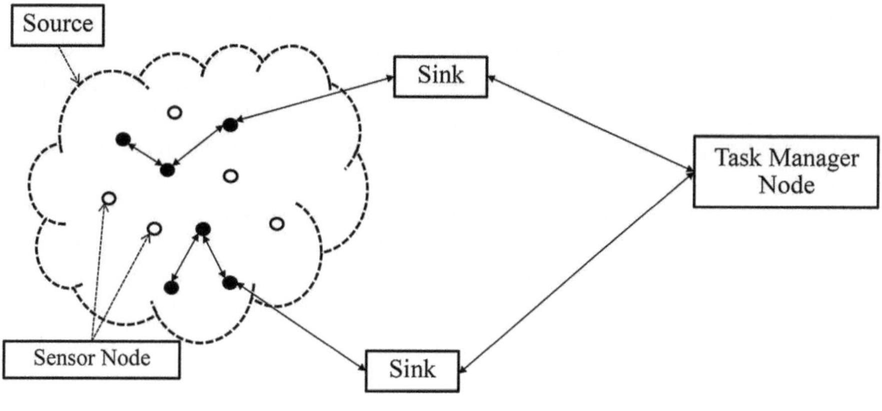

Fig. 4.6 WSNs communication architecture

military, the remote deployment of sensors is used for monitoring movements of the enemy troop. For traffic management and surveillance, the sensors embedded in the roads are used to monitor traffic flows and provide real-time route updating. For the future cars, multiple wireless sensors are expected to handle the accidents or even the thefts. For healthcare monitoring, WSNs is used to collect different types of the clinical data of the chronic patients for making the decision whether the patient requires specific care. Currently, the application in healthcare requires specific types of sensors that can be attached to the body obtrusively and communicate among all of the others. This requirement enables new type of the communication network as Body Area Network (BAN).

4.3 Body Area Networks

For current context-aware applications, much interest goes to healthcare oriented applications requiring clinical data from the patient's body. It is evitable that the body sensors have been gaining interest to fulfill these dramatically demands. Therefore, the networks as Body Area Networks (BANs) for facilitating the communication among these small and tiny sensors are required. Nowadays, the concept of BANs which is also called Body Sensor Networks (BSNs) is widely introduced primarily in medical and healthcare application domains. BAN technology emerges as the natural by-product of existing sensor network technology and biomedical engineering (Karulf 2008). BANs are firstly introduced as the subgroup of PANs (Yang and Yacoub 2006). As mentioned before, PAN is the interconnection of information technology devices within the range of a person normally within 10 meters. On the other hand, BANs are the combination of small intelligent devices which are attached or implanted in the body. Those devices are capable of wireless communication. A wireless BAN (WBAN) is formally defined by IEEE 802.15[1] as "a communication standard optimized for low power devices and operation on, in or around the human body (but not limited to humans) to serve a variety of applications including medical, consumer electronics/personal entertainment and other". When the WPAN working group realizes that there was the need for a standard for using the devices inside and close to the human body, more suitable standard is established. Consequently, IEEE 802.15.6[2] is a new standard for WBANs and has many advantages over other wireless communication standards (Kwak et al. 2010).

With some specific requirements, WBANs require some concrete supports that cannot obtain from the existing standard. For example, IEEE 802.11 is the standard group for WLAN or Wi-Fi, including different usage areas such as IEEE 802.11a/b/g/n. Since WLANs are mainly used by the computers or the portable

[1]http://standards.ieee.org/about/get/802/802.15.html.

[2]http://standards.ieee.org/findstds/standard/802.15.6-2012.html.

devices, they do not concern much about the power consumption. IEEE 802.15 standard focusing on short range, low complexity, cheap and tiny power consumption is mainly designed for WPAN. The familiar examples are such as IEEE 802.15.1 for Bluetooth and IEEE 802.15.4 for Zigbee. These technologies are just only about the human body not on the human body. For this reason, IEEE 802.15.6 has been developed. The construction of WBANs needs to consider some important concerns such as energy consumption, quality of service (QoS), co-existence, security, and privacy, etc. For example, the battery of WBAN sensor needs to ensure the sufficient and lifelong energy consumption. For QoS and reliability, WBANs should be able to transmit error-free data in real time. For co-existence, WBANs should be able to operate across different networks without any interference. For security and privacy, WBANs can have very crucial information, so it has to be ensured that the security and privacy are configured appropriately.

Basically, IEEE 802.15.6 standard is presented within three layers including physical layer (PHY), medium access layer (MAC), and security layer (Kwak et al. 2010). Since existing MAC protocols offered for WLAN, Bluetooth, ZigBee are not satisfied with the reliable low-power transmission, IEEE 802.15.6 provides a new MAC layer satisfying the reliable low-power transmission. For the secure communication, IEEE 802.15.6 defines three levels including Level 0 for unsecured communication, Level 1 for authentication only, and Level 2 both authentication and encryption. The brief specification of WBAN (Karulf 2008) is shown in Table 4.4.

For WBAN, there are three devices physically attached to human body including sensors, actuators, and personal devices. The characteristics of WBAN sensors are special designed because it is attached to the human body. The precise of the sensors is required, but the safety for the human body is very important. WBAN sensors have several features that make them suitable for being used in many applications. For example, the sensors are designed to handle the power resources optionally so that the nodes remain alive after the long lifetime of applications. The sensors are designed to deal with heterogeneous of sensed data. Besides detecting different data such as pulse rate, blood pressure, heart rate, etc., the sensors may need to address different kinds of capacity, computation capability, energy consumption, etc. Finally, the sensors need to be straightforward and profitable so that they can be easily carried out and afforded. Some available commercial sensors are Electrocardiography (ECG) sensor, blood pressure sensors, carbon dioxide (CO_2) gas sensor, humidity sensors, temperature sensors, Electroencephalography (EEG) sensors, etc. For the actuator, the hardware architecture of the actuator node is similar to the sensor, but it has an additional hardware called the actuator hardware to take any responded actions. Finally, personal devices collect information which has been gathered by the sensors and transmit this information to other devices, users or actuators.

For communication network, WBAN is composed of one or more Body Sensor Units (BSU), one Body Central Unit (BCU) and a link with other long range networks (Latré et al. 2011; Malik and Singh 2013). Different BSUs collect different information about human bodies such as respiration rate, pulse, glucose rate,

Table 4.4 Brief Specification of WBANs

Characteristic	IEEE 802.15.6 Detail
Configuration	Single scalable MAC
Power consumption	Very low power for human tissue
Power source	Compatible with body energy operation
Quality of Service (QoS)	Reliable response to external stimuli
Frequency band	Regulatory and/or medical authority approved band
Channel	Air, around and inside human body
Safety	Required
Application	Medical, entertainment, gaming, sport, etc.
Rang	2–5 m
Data rate	1 Kbps–10 Mbps
Power consumption	0.01 mW standard model
Network size	<256 devices per band

ECG, etc. Then, each BSU sends its data to the BCU. BANs communication conceptual diagram (Isikman et al. 2011) is shown in Fig. 4.7.

For BANs or BSNs, three different infrastructures (Karulf 2008) are used including Managed Body Sensor Networks (MBSN), Autonomous Body Sensor Network (ABSN) and Intelligent Body Sensor Networks (IBSN). For MBSN, BCU sends an alert message to the closest hospital or the doctor after a problem has been detected through BSU. The third person is required to make a decision and send it back to the BCU. The actuators will execute the decided response. The network needs to be linked with a long range connection such as mobile network, Wireless Fidelity (WIFI) to the internet, or another network to accommodate the decision process. Extra-BAN (EBAN) communication which can be any technology supporting wireless broadband data transfer and are not belong to BAN such as WPAN, WLAN, etc. are needed. For ABSN, the BCU is more intelligent and has been trained to make the decision by itself. The objective of this structure is to use ABNS to monitor and protect the patients autonomously. Therefore, the BCU can analyze the different inputs, perform a diagnosis and give orders to the actuators to take any action. The actuators are also attached to the body. For example, a body actuator can inject insulin when the BSU measure the critical glucose rate of the user without any intervention from the third person. Lastly, IBSN is a combination of both networks. As a result, the decision can be made by actuator node but it will be sent to the third person to make the decision for more complex decisions.

WBANs have initially gained the motivation from health monitoring and prevention. Ubiquitous healthcare is one of the popular context-aware applications using WBANs. It is an emerging technology that promises increasing in efficiency, accuracy, and availability of medical treatment. Current examples of WBANs applications are in sports and fitness, military, emergency services, emotion detection, personal health monitoring, etc.

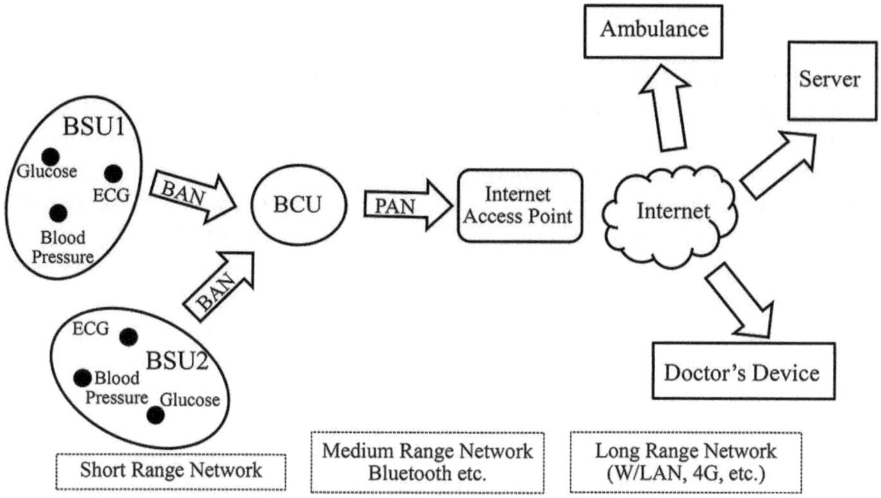

Fig. 4.7 BANs communication conceptual diagram

4.4 Social Networks

Currently, billions of users participate in social networks to form communities, to produce, and to consume media contents in different ways. As a result, the social network has become very crucial for context-aware applications nowadays. Accordingly to the concept of the social-aware application, the social context derived from social network information is the primary context for enabling the appropriate response to the group of users not only for the individual user. Social networks provide a compelling description of the social structure and support the dynamic interaction among people for the current society. These interactions affect great contributions through social-aware applications. From Wikipedia definition, a social network is "a social structure made of individuals (or organizations) called nodes which are tied (connected) by one or more specific types of interdependency, such as friendship, kinship, financial exchange, dislike or relationships of beliefs, knowledge or prestige." The social network is also introduced together with its measurements as social network analysis (Wasserman and Faust 1994). Like a computer network which is a set of machines connected by different types of communication methods, a social network thus is a set of people who are linked by various types of social relationships such as mentorship, friendship, co-working, or information exchange (Wasserman and Faust 1994). This definition is widely used for defining the definition of social context currently. In another word, a social network describes the social structure between the actors who can be mostly the individuals or the organizations. It simply shows the ways in which people are connected with different kinds of relationships. At the same time, the social network is studied to determine the characteristic of the network which connects people together.

Social networks can be analyzed concerning egocentric networks and sociometric networks. The egocentric networks are also known as personal networks or ego networks (Nam et al. 2015). Ego means the person at the center together with alters or other members in the network who are directly connected to the ego (Borgatti et al. 2013). Egocentric data can be collected by using the question concerning the phenomena affecting individuals to index person. More specifically, the egocentric data is widely used for capturing social influence and social support of that single individual (Valente 2010). Additional information can also be collected on the relationships between the respondent and all respondents' attributes such as demographic, relationships, behaviors, etc. For sociometric networks, the whole network composes of the network of the networks at the level of communities (Borgatti et al. 2013). Sociometric networks aim to assess the collective dimension of social ties which are also called the web of relationships (Valente 2010). The sociometric data requires interviewing all people within the community of interest (Wasserman and Faust 1994). Also, the sociometric networks are particularly useful for studying the dynamic changes in network structure over time.

Social networks are commonly used by the online community of internet users having common interests in hobbies, religion, or politics, etc. through social network sites. They want to socialize on the sites by reading the profile pages of other members and possibly even contacting them. Social network sites are used nowadays in many different domains. They are designed and developed to allow individuals to present themselves particularly to their social networks and establish or maintain connections with the others. Currently, some popular social network sites are such as professional/work related site as LinkedIn,[3] microblogging site as Twitter,[4] romantic relationship related site as Friendster,[5] personal related site as Facebook,[6] music or politics related site as Myspace,[7] photos/picture related site as flickretc,[8] etc. The critical studies of social networks are to determine their characteristics and definitely to determine social contexts. As a result, social network analysis is mainly introduced to determine social context for any social-aware application.

---

[3]www.LinkedIn.com.

[4]www.twitter.com.

[5]www.Friendster.com.

[6]www.facebook.com.

[7]www.MySpace.com.

[8]www.flickretc.com.

4.4.1 Social Network Analysis

Social network analysis (SNA) (Wasserman and Faust 1994) is one of the popular methods for measuring social context. It is the method for mapping and measuring of relationships between computer, people, groups, organizations, or other information processing entities. More specifically, SNA can be considered as a set of formal analytic tools gaining much attention across many application domains (Edwards 2010) such as sociology, anthropology, economics, politics, psychology, business, mathematics, etc. (Freeman 2004). The approach to analyse the social networks can be qualitative, quantitative and mixed methodologies. The methodologies also mean both data collection and analysis. The quantitative-based methodology (Fischer 1982; Wellman 1979) can map and measure social relation in a systematic way, while the qualitative-based methodology (Barnes 1954; Trotter 1999) can better reveal process, change, content, and context (Edwards 2010). The mixed based method (Mønsted 1995; Jack 2010; Knox et al. 2006) is the combination of quantitative-based and qualitative-based methodologies for both data collecting and analyzing of rational data. In this section, the quantitative-based methodology is mainly introduced.

The fundamental unit of SNA is the relation, which is characterized by content, direction, and strength. Content means the resource that is exchanged, while direction means the type of communication which can be directed or undirected. Strength means the frequency or the volume of communications. In the past, gathering social network data involves both the observation and the record of activities. The general ways of social network data gathering method can be easily done by using questionnaires, interviews, diaries, etc. Even though the required data can be obtained, its reliability is often questioned. Incorrect reporting, either intentionally or not, often occurs when the participants record their activities (Temdee and Korba 2001). For instance, the received data may be biased because not all interactions are well remembered. Therefore, nowadays automatic social network gathering is commonly used to overcome these disadvantages, but the overwhelming social data may need the complicated data processing.

4.4.2 Graph Theory for Social Network

Graph theory provides the unifying language for network structure (Bondy and Murty 1976). It has been playing the leading roles to explain the connections between entities in many application domains, especially in the social network. Graph-based representation is the representation of a problem as a graph topology. It can represent the problem with different point of view and can make a problem much easier to solve. Therefore, the solution can be more accurate if the appropriate solving tools are given. The graph-like networks can be found in general such as friendship network, academic collaboration network, business network, protein

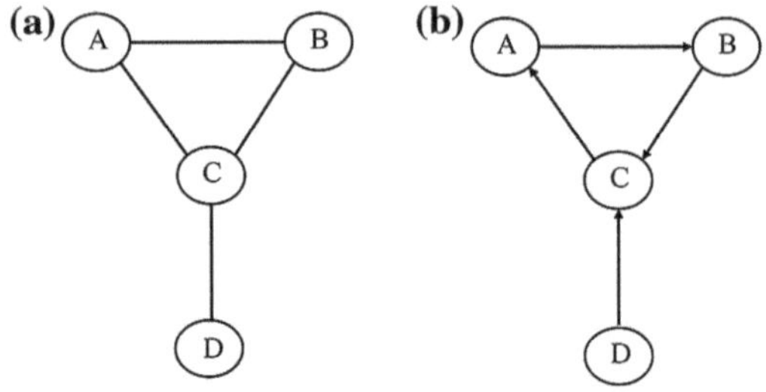

Fig. 4.8 Undirected and directed graphs

interaction network, transportation network, Internet, ecological network, etc. This section thus introduces basic graph theory necessary for better understanding of social network and its analysis. The concepts explained here are mostly from the works of Alddous and Wassan (Aldous and Wilson 2003; Wasserman and Faust 1994).

More specifically, a graph or a network is a way to specify relationships amongst a collection of items. A graph normally consists of the set of objects called nodes and the pairs of objects called edges. Two nodes are neighbors if they are linked by an edge. Figure 4.8 shows examples of graphs that both of them have four nodes including A, B, C and D. As seen from Fig. 4.8a, this graph is undirected since the edges have no orientation with default assumption. At the same time, the graph in Fig. 4.8b is directed graph since the edges have a direction including the edge from A to B, B to C, C to A and D to C.

Graph Structures

Graph structures are used to show the interesting and relevant sections of a graph by using the structural metrics to explain a structural property. There are two kinds of metrics including global and local metrics. While the global metrics refer to a whole graph, the local metrics refer to only a single node in a graph. Identify interesting sections of a graph is challenging because they are domain-specific structure. A graph is connected if any two nodes are connected by a path or one node can get to one node by following a sequence of edges.

Edges

The edges sometimes can carry additional information. For example, the signs can represent the positive and negative attitude such as friends or enemies. The tie strength can describe the degree of relationship such as friendship, mentorship, etc. The distance can explicitly represent the distances between nodes such as how far between cities on the map. The delay can represent the time consumptions such as how long the transmission takes the signal between nodes. In conclusion, the information carried by edges is domain specific. Figure 4.9 shows the edge list of the graph.

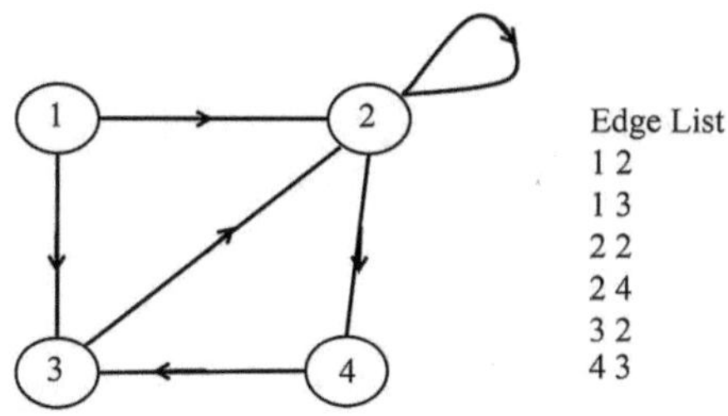

Fig. 4.9 Edge list of the graph

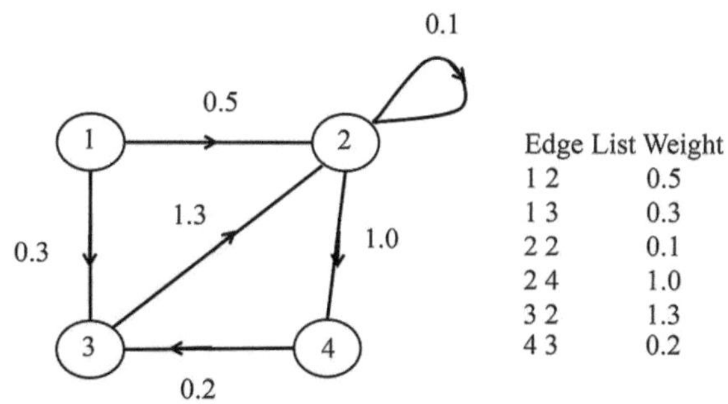

Fig. 4.10 Edge list for weighted graph

For a weighted graph, every edge has an associated number called a weight. At the same time, every edge has a + or a – sign associated with it in a signed graph. Figure 4.10 shows edge list with its weight of the graph.

Walks

A walk of length k in a graph is a succession of k edges which are not necessarily different in the form uv, vw, wx, …, yz. This form is called a walk between u and z. The walk is close if u = z. Consequently, this walk is called close walk. Examples of walk and close walk are shown in Fig. 4.11.

Paths between nodes

A path is a walk in which all the edges and all the nodes are different. More specifically, a path is the sequence of nodes with the property that an edge connects each consecutive pair in the sequence. It can also be defined as a sequence of edges. Figure 4.11 also shows the examples of walks and paths in the undirected graph.

Cycle

A closed path where the edges are all different is called cycle. The examples of cycles are shown in Fig. 4.12.

Degree

For the undirected graph, the degree is the number of edges incident on a node. For the directed graph, there are two type of degree including in-degree and out-degree.

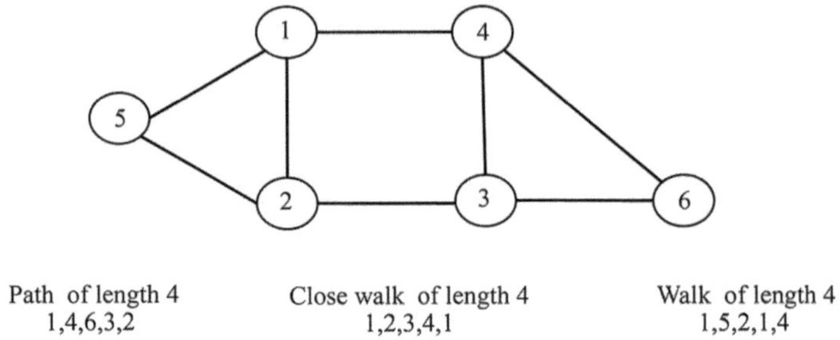

Fig. 4.11 Examples of walks and paths for undirected graph

Fig. 4.12 Examples of cycles for undirected graph

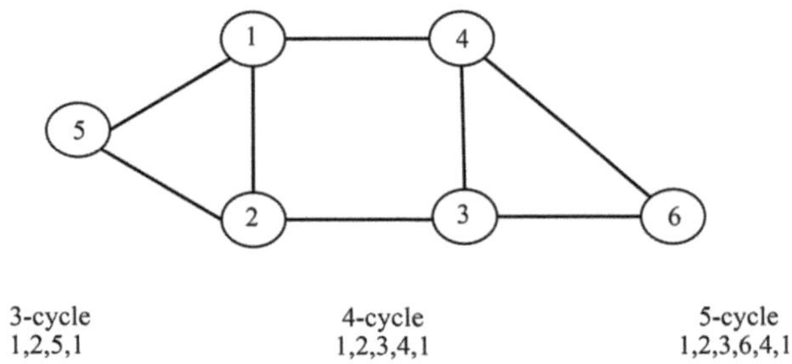

In-degree is the number of edges entering while out-degree is the number of edges leaving. Figure 4.13 shows degree for undirected graph and Fig. 4.14 shows in-degree and out-degree of directed graph respectively.

Special Types of Graphs

Some particular graphs are general found in many application domains and have specific characteristics. For example, the empty graph or edgeless is simply defined as the graph with no edge as shown in Fig. 4.15. Also, the null graph is the graph with no nodes and no edge. Tree graph is connected acyclic graph in which two nodes have exactly one path between them. Figure 4.16 shows the examples of tree

Fig. 4.13 Degree for undirected graph

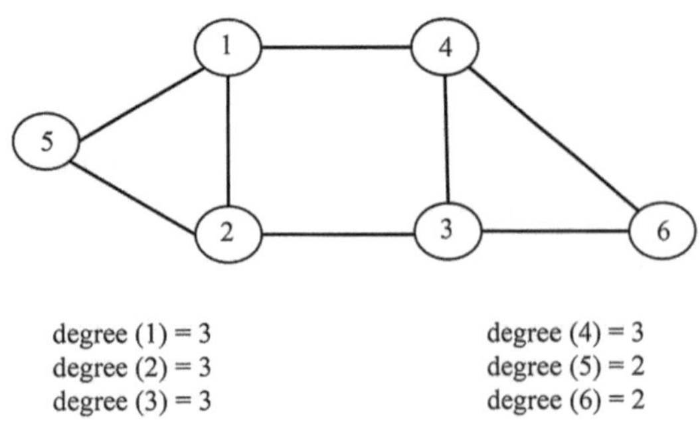

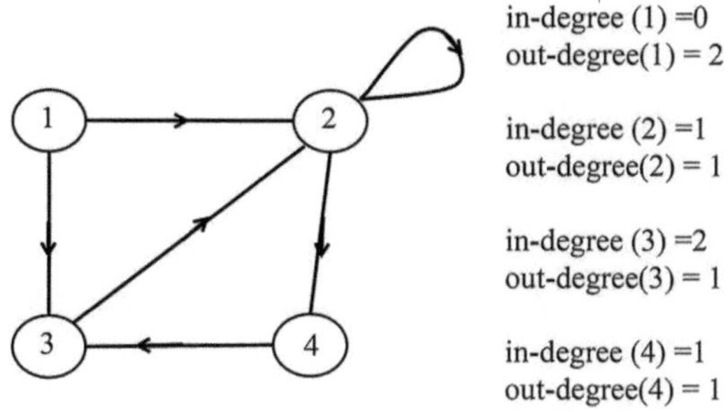

Fig. 4.14 In-degree and out-degree for directed graph

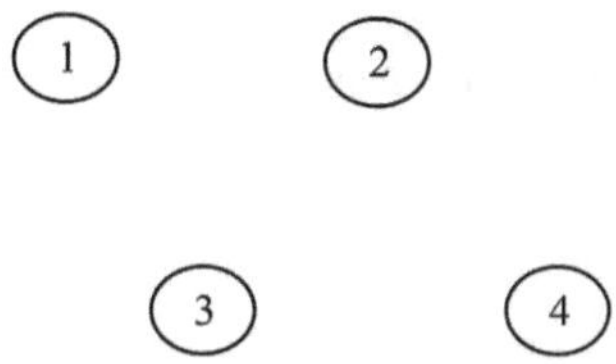

Fig. 4.15 Empty graph and null graph

graphs. The special tree graphs which are path and star are shown in Fig. 4.17a, b respectively.

A regular graph is the connected graph that all nodes have the same degree as shown in Fig. 4.18.

Next section, social network analysis is introduced in detail in term of its definition, measurements, and implementation for the real social network application.

4.4.3 Social Network Analysis Measurements

Social Network Measurement (SNA) (Wasserman and Faust 1994) is the measuring and the mapping of relationships between any social entities such as people, group, organizations, etc. The nodes represent the social entities while the links represent the relationships or information flows. SNA provides two ways of data representation methods including mathematical and graphical methods. Three mathematical foundations are generally found to explain this measurement method including graph theory, statistical and probability theory, and algebraic models. The Sociogram is used for illustrating such relationships. For Sociogram, the social entities are represented as the points in two-dimensional space and the relationships among pairs of them are represented by the lines linking the corresponding points. The example of Sociogram is shown in Fig. 4.19. From Fig. 4.19, there are five actors or members connected. The nodes represent the members, while the links show the communication among all members. This Sociogram represents undirected communication of those members. The strength indicated by the number represents the volume of communications.

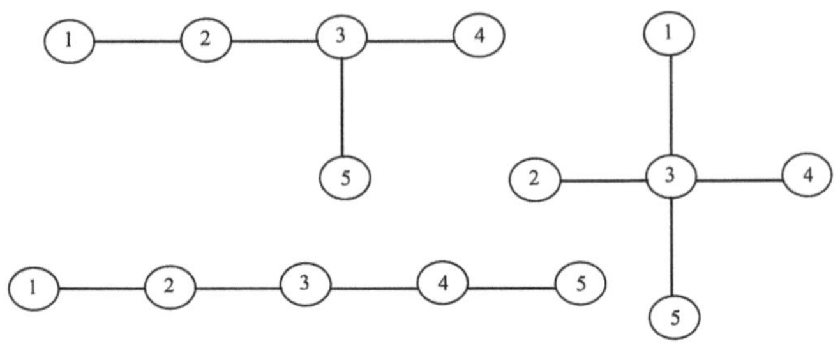

Fig. 4.16 Examples of tree graphs

Fig. 4.17 Path and star

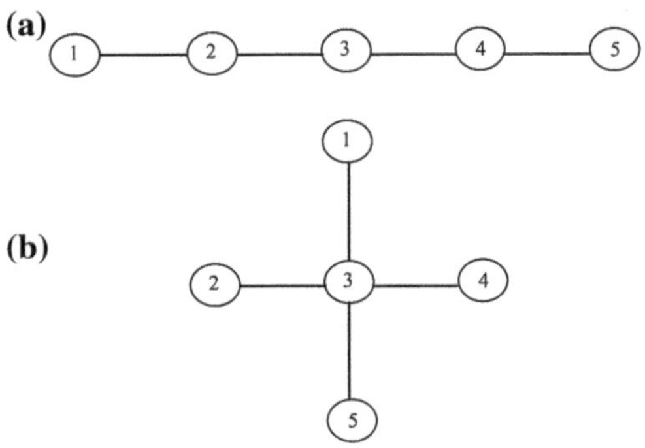

Fig. 4.18 Regular graph

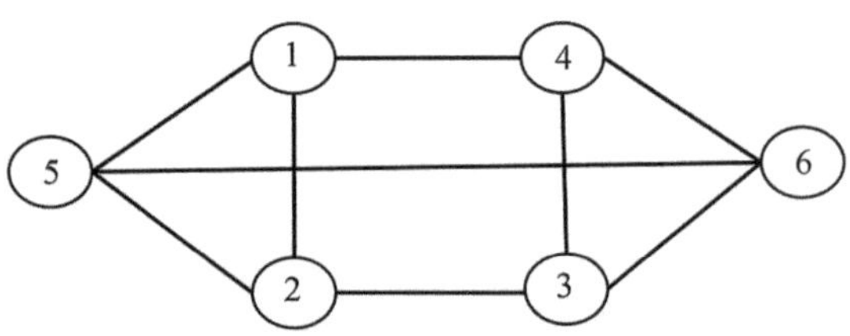

From the Sociogram, two members are connected because they talk to each other, or interact in some ways. As shown in Fig. 4.19, A regularly interacts with C, but not with D. Therefore, A and C are connected, while there is no link drawn between A and D. All connections can be expressed in the matrix called Sociometrix as shown in Fig. 4.20.

From the Sociometrix, the numbers represent the direct connections between each pair of actors. It is shown in Fig. 4.20 that D and E have the maximum connections while there are many pairs of members having no direct connection to each other at all such as A and B, A and D, A and E, etc. In conclusion, Sociogram and Sociometrix represent the same information with different ways to portray.

SNA provides the mathematical model to describe and analyze the network position of individuals in the social network. Network centrality is used to

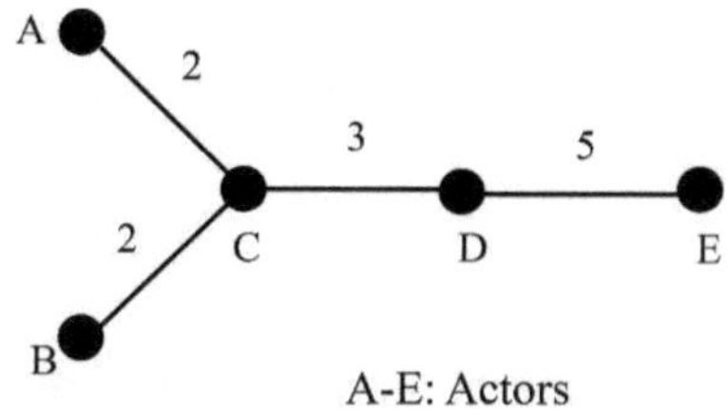

Fig. 4.19 Example of sociogram

Fig. 4.20 Sociometric

	A	B	C	D	E
A	0	0	2	0	0
B	0	0	2	0	0
C	2	2	0	3	0
D	0	0	3	0	5
E	0	0	0	5	0

determine which nodes are most central based on different contexts and purposes. Finding out which node is the most central is important because this node can help disseminating the information faster, stopping epidemics, protecting the network from breaking, etc. The popular network centralities are introduced in this section including the degree of centrality, closeness centrality, and betweenness centrality.

Degree of Centrality

Accordingly to Freeman's work (Freeman 2004), the best centrality from the number of connections can identify some people who are the most popular in the network. These nodes are the people whom most people in the network would like to talk to. At the same time, these nodes can also be the people who do favors for other people. As it can be seen from Fig. 4.21, the middle node has the maximum degree of centrality. From Fig. 4.21a, X has more in-degree of centrality than Y. It can be concluded that X is more popular than Y. X can be considered as the person whom most people in the network would like to talk to. At the same time, X in Fig. 4.21b also has more out-degree of centrality than Y. It can be concluded that X is more popular than Y. In addition, X is the person who mostly does favors for other people in the network.

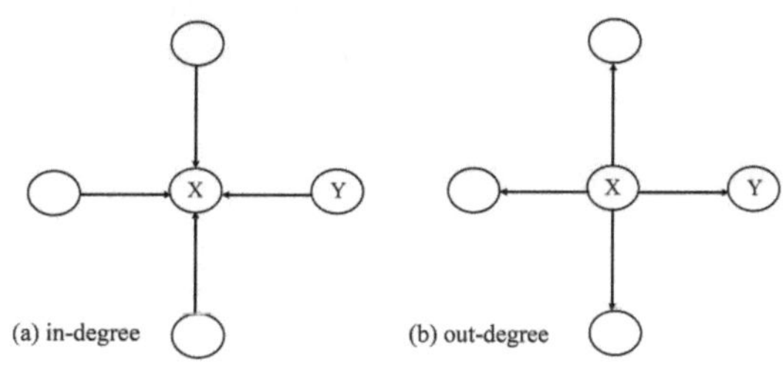

Fig. 4.21 Degree of centrality

The normalized degree of centrality is shown in (4.1), where g is the total number of nodes and $C_D(n^*)$ is the maximum degree of centrality in the network.

$$C_D' = \frac{\sum_{i=1}^{g} [C_D(n*) - C_D(n_i)]}{[(g-1)(g-2)]} \quad (4.1)$$

From Fig. 4.19, it can be seen that D has the most direct connections in the network, making D as the most active actor in the network. In other words, D is the star of the network. It is more likely that D is the most important member but not always.

Closeness Centrality

Closeness represents the people in the network who are the one in the middle of the networks. Although they are not so important to have many direct friends, they are not too far from the center. More specifically, closeness focuses on how close an actor is to all other actors in the network. The actor is central if it can quickly interact with all others. If the actors in the set are engaged in problem-solving, the efficient solutions may occur when the actor has very short communication paths to others.

Closeness can also be calculated as a measurement of the distribution unbalancing of distances across the actors. These measurements depend on the summation of the geodesic distances from each actor to all others. Closeness is defined as the length of the average shortest path between a vertex and all vertices in the graph as shown in (4.2).

$$C_C(n_i) = \frac{1}{\sum_{j=1}^{g} d(n_i, n_j)} \quad (4.2)$$

where $d(n_i, n_j)$ is the length of shortest path between actor n_i and actor n_j or the numbers of step for actor n_i need to reach actor n_j. The normalized closeness is defined as shown in (4.3).

$$C_C' = (g-1)C_C(n_i) \quad (4.3)$$

From Fig. 4.22, X has more closeness than Y because X is located relatively in the middle of the network.

From Fig. 4.19, C has maximum closeness. It means that C has the shortest paths to all others or C is close to everyone else. C is in the excellent position to monitor the information flow in the network or to know what is happening in the network.

Betweenness Centrality

Betweenness presents the people in the networks that many pairs of individuals would have to go through to reach one another in the minimum number of hops. These people act as brokers between groups. There is the likelihood that information originating anywhere in the network will reach these people. From

Fig. 4.22 Closeness centrality

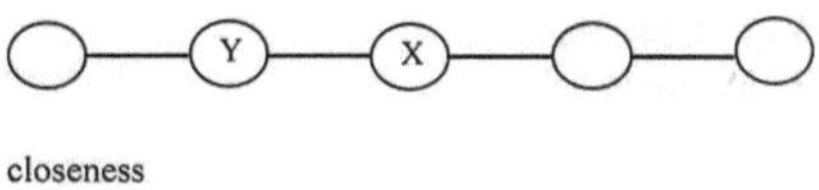

Fig. 4.23, X has more betweenness centrality than Y. It can be seen that X acts as a bridge between two smaller groups in the network.

As Betweenness is identified as the actor lies on several paths among other pairs of actors, such actor has the control over the flow of information in the network. It is the sum of probabilities across all possible pairs of actors, that the shortest path between y and z will pass through actor x.

$$C_B(n_i) = \sum_{j<k} \frac{g_{jk}(n_i)}{g_{jk}} \tag{4.4}$$

where $g_{jk}(n_i)$ is the numbers of shortest paths between n_j and n_k through n_i and g_{jk} is the numbers of shortest paths between n_j and n_k. The normalized version for the undirected network is shown in (4.5), and the normalized version for the directed network is shown in (4.6) respectively.

$$C'_B(n_i) = \frac{C_B(n_i)}{(g-1)(g-2)/2} \tag{4.5}$$

$$C'_B(n_i) = \frac{C_B(n_i)}{(g-1)(g-2)} \tag{4.6}$$

Figure 4.24 shows the examples how to count the numbers of pairs of the individual. It can be seen that A lies between no pair of other nodes. B lies between A and other nodes including C, D, and E. C lies between (A, D), (A, E), (B, D), (B, E). It can be seen that C is only one alternative for all pairs to be connected.

From Fig. 4.19, C has the maximum betweenness. Generally speaking, C is between two important small groups. C plays a broker role in the network. However, C is also a single point of failure. Without C, two pairs of the team would be separated. Therefore, an actor with high betweenness has the significant influence over what flows in the network. In conclusion, D has the maximum degree of

Fig. 4.23 Betweenness centrality

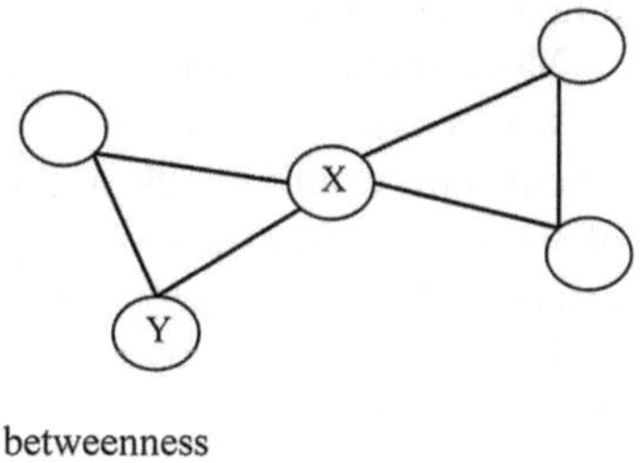

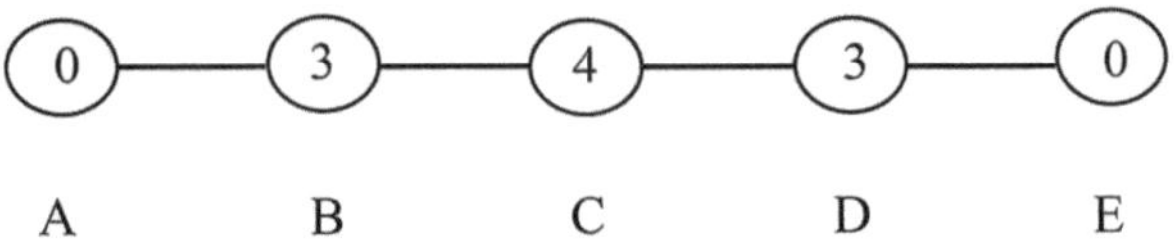

Fig. 4.24 Pairs of individuals

centrality but not maximum closeness and betweenness. C has the maximum closeness and betweenness. It can be concluded that the member who is the most active person in the team is not always the person who is relatively close and has the most interpersonal influence to anybody else in the team.

Bonachich Power Centrality

Bonachich Power Centrality is the centrality measurement when one's centrality depends on neighbors' centrality. Bonacich Power Centrality employs an iterative estimation approach which gives the weights to each node's centrality by considering the centrality of the other nodes to which it is connected. Therefore, one's centrality depends also on the connections of its neighbors. It can be seen that the actors who are connected with the very central neighbors should have more centrality or prestige than those who are not. The attenuation factor or the weight is used for calculating of the Bonacich Power measurement. For positive attenuation factor (between 0 and 1), it means that one's power is enhanced by being connected to well-connected neighbors. Alternatively, the actors who are well connected to not well-connected individuals are probably powerful because the others are dependent on them. For a negative attenuation factor (between 0 and -1), it can be used for computing the proper power. The Bonachich Power Centrality (Cullen et al. 2015) can be calculated from (4.7).

$$C(\alpha, \beta) = \alpha(I - \beta R)^{-1} R1 \tag{4.7}$$

where α is a scaling vector which is set to normalize the score, β represents how the centrality of people ego is tied to be weighted, R is the adjacency matrix, I is the identity matrix and 1 is a matrix of all ones. The magnitude of β represents the radius of power. While the small value weights local structure, the larger value weights global structure. If $\beta > 0$, ego has higher centrality when conncted to people who are central. If $\beta < 0$, ego has higher centrality when connected to people who are not central. Finally, when $\beta = 0$, degree centrality of eco can be determined. Figure 4.25 shows different values of Bonachich Power Centrality with different β values.

More social network measurements are used for explaining different types of networks and identifying different roles of the actors in the networks such as eigenvector centrality, clustering coefficient, cohesion, integration, reach, etc. Each of these measurements is chosen dependently to the application's purposes. For example, the eigenvector centrality is frequently used to identify the most critical node of the networks by evaluating the relative scores assigned to all nodes in the network. Clustering coefficient is a measure of the possibility that two nodes are

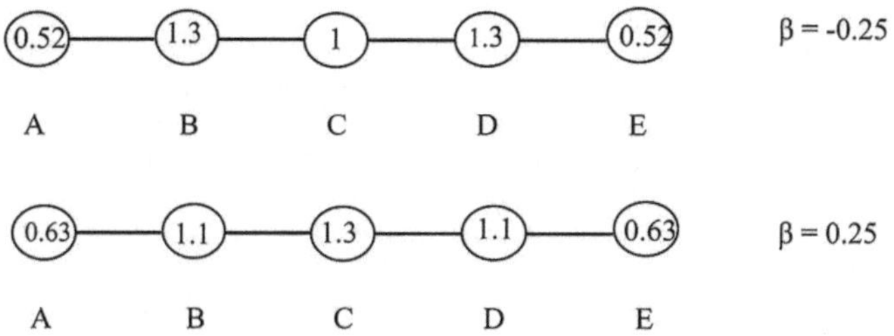

Fig. 4.25 Bonachich power centrality with different β values

associated by themselves or not. At the same time, the higher clustering coefficient indicates a larger cliquishness. Cohesion refers to the degree that the actors are connected directly to each other by cohesive bonds. Groups are identified as cliques if every actor is directly linked to every other actor. As mentioned before that selecting social network measurement mainly depends on specific domain applications, the combination of social network measurements is commonly used in many social-aware applications. For example, the group context is mostly used in the social-aware application where it is used for initiating the appropriate response to the group. Group context can be combined with many social network analysis measurements such as cohesion, integration, and reach. Also, social network analysis is still suitable to explain the social interactions of any entity which does not particularly mean only for the people. It is still applicable to understanding the social interaction among machines or between the machine and the people which will be able to provide more useful applications for context-aware computing in the future.

This chapter has introduced the reader about some possible communication networks have been widely used for existing context-aware applications. Moreover, this chapter also points out some communication networks which can be generally found in existing applications. However, the crucial issue regarding the communication is the security. More detail and some concerns regarding the security will be explained and discussed in the next chapter.

References

Aldous, J. M., & Wilson, R. J. (2003). *Graphs and applications: An introductory approach* (Vol. 1). Berlin: Springer Science & Business Media.

Arjmandi, M. K. (2016). 5G Overview: Key Technologies. In *Opportunities in 5G Networks: A Research and Development Perspective* (pp. 19–32). CRC Press.

Barnes, J. A. (1954). *Class and committees in a Norwegian island parish*. New York: Plenum.

Bondy, J. A., & Murty, U. S. R. (1976). *Graph theory with applications* (Vol. 290). London: Macmillan.

Borgatti, S. P., Everett, M. G., & Johnson, J. C. (2013). *Analyzing social networks*. New York: SAGE Publications Limited.

Cullen, K. L., Gerbasi, A., & Chrobot-Mason, D. (2015). Thriving in central network positions: The role of political skill. *Journal of Management,* 0149206315571154.

Edwards, G. (2010). *Mixed-method approaches to social network analysis.*

Estrin, D., Girod, L., Pottie, G., & Srivastava, M. (2001). Instrumenting the world with wireless sensor networks. In *Proceedings. (ICASSP'01). 2001 IEEE International Conference on Acoustics, Speech, and Signal Processing, 2001* (Vol. 4, pp. 2033–2036). IEEE.

Fischer, C. S. (1982). *To dwell among friends: Personal networks in town and city.* Chicago: University of Chicago Press.

Freeman, L. (2004). *The development of social network analysis.* A Study in the Sociology of Science.

Goldsmith, A. (2005). *Wireless communications.* Cambridge university press.

Hussain, M., Gawate, S. P., Prasad, P. S., & Kamble, P. A. (2015, April). Smart irrigation system with three level access mechanisms. In *Computation of Power, Energy Information and Commuincation (ICCPEIC), 2015 International Conference on* (pp. 0269–0275). IEEE.

Isikman, A. O., Cazalon, L., Chen, F., & Li, P. (2011). Body area networks. *Mobile Networks and Applications Journal, 16*(2), 171–193.

Jack, S. L. (2010). Approaches to studying networks: Implications and outcomes. *Journal of Business Venturing, 25*(1), 120–137.

Kao, C. H. (2002). *Performance of the IEEE 802.11 a wireless LAN standard over frequency-selective, slow, ricean fading channels.* Monterey, CA: Naval Postgraduate School

Karulf, E. (2008). *Body area networks (BAN).* April 23, 2008 [2013-12-22]. http://www.cse.wustl.edu/~jain/cse574-08/ftp/ban/index.html

Knox, H., Savage, M., & Harvey, P. (2006). Social networks and the study of relations: Networks as method, metaphor and form. *Economy and Society, 35*(1), 113–140.

Kwak, K. S., Ullah, S., & Ullah, N. (2010, November). An overview of IEEE 802.15. 6 standard. In *2010 3rd International Symposium on Applied Sciences in Biomedical and Communication Technologies (ISABEL 2010)* (pp. 1–6). IEEE.

Latré, B., Braem, B., Moerman, I., Blondia, C., & Demeester, P. (2011). A survey on wireless body area networks. *Wireless Networks, 17*(1), 1–18.

Malik, B., & Singh, V. R. (2013). A survey of research in WBAN for biomedical and scientific applications. *Health and Technology, 3*(3), 227–235.

Mønsted, M. (1995). Processes and structures of networks: Reflections on methodology. *Entrepreneurship & Regional Development, 7*(3), 193–214.

Nam, S., Redeker, N., & Whittemore, R. (2015). Social networks and future direction for obesity research: A scoping review. *Nursing Outlook, 63*(3), 299–317.

Temdee, P., & Korba, L. (2001). Of networks, interactions and agents: an approach for social network analysis. In *Computer Supported Cooperative Work in Design, The Sixth International Conference on, 2001* (pp. 324–329). IEEE.

Trotter, R. (1999). Friends, relatives and relevant others: Conducting ethnographic network studies. In J. J. Schensul et al (Eds.), *Mapping social networks, spatial data, and hidden populations.* London, Altamira.

Valente, T. W. (2010). *Social networks and health: Models, methods, and applications.* Oxford: Oxford University Press.

Wasserman, S., & Faust, K. (1994). *Social network analysis: Methods and applications* (Vol. 8). Cambridge: Cambridge University Press.

Wellman, B. (1979). The community question: The intimate networks of East Yorkers. *American Journal of Sociology,* 1201–1231.

Wood, A. D., Stankovic, J. A., Virone, G., Selavo, L., He, Z., Cao, Q., … & Stoleru, R. (2008). Context-aware wireless sensor networks for assisted living and residential monitoring. *IEEE Network, 22*(4), 26–33.

Yang, G. Z., & Yacoub, M. (2006). *Body sensor networks.*

Chapter 5
Security for Context-Aware Applications

Abstract With the advance of wireless technology, the contexts are transmitted through fast and efficient communication method. All computing activities occur around the users obtrusively. To satisfy context-aware applications, it is tough to maximize functionality but remains strong security at the same time. This chapter introduces the principle concept of security required for context-aware applications. The security, in general, is firstly introduced. Some security attacks and countermeasures are reported. Some security recommendations for context-aware applications and the existing security protocols are also discussed in this chapter. The ultimate goal of this chapter is to promote sense of security awareness for the readers who can be both the developers and the users.

Security is very crucial for any application including context-aware applications because the personal information is gathered and utilized autonomously with less conspicuous. With the advance of wireless technology, the contexts are transmitted through fast and efficient communication method. All computing activities occur around the users obtrusively. The current context-aware applications present good examples of the needs of security, especially for IoT applications. For example, the connected medical devices are allowed by the patients to work with their caregivers to manage their diseases. At home, the energy consumption can be controlled by the collaboration between the smart meter and several electrical appliances. In the car, the driver can be worn about the traffic and road condition. The users agree that IoT will offer numerous and potentially revolutionary benefiting to them. However, the users should notice some risks from these convenient services. For example, the users should be aware that IoT can present a variety of potential security risks to abuse the users and their assets in many ways. More specifically, IoT applications can enable unauthorized access and misuse of personal information without the appropriate prevention. They can even facilitate attacks on other systems conveniently. Moreover, the privacy of the users may flow from the collection of personal information, habits, locations, and physical conditions over time. This personal

P. Temdee and R. Prasad, *Context-Aware Communication and Computing: Applications for Smart Environment*, Springer Series in Wireless Technology,
DOI 10.1007/978-3-319-59035-6_5

information usually benefits third parties. Although there is the agreement that the IoT products have to provide reasonable security mechanism, the users still need be aware of their own security. Therefore, it is very crucial to pay critical concern on security related issues. This chapter introduces the principle of security, particularly for context-aware applications. More detail of some security issues from relevant communication channels is discussed. The reader thus can get the idea of the threats and the current protection mechanisms. Additionally, the reader should be able to suggest the basic requirement of security issue for their applications after completing this chapter.

5.1 Security in General

Security has three different types including computer security, network security, and Internet security. Computer security can be considered as the tools for protecting data and defending the hacker at the same time. Network security focuses on protecting the data during the transmission. Internet security also focuses on protecting the data during the transmission especially over a collection of interconnected networks. In this chapter, network security and Internet security are mainly focused. In the past, the security is a young and immature field. The attackers are usually more innovative than the defenders who are usually in fear, uncertainty, and doubt. The back attacking is still illegal. Currently, the security becomes a scientific discipline which normally is the application and the technology centric. At the same time, the back attacking will be an integral part of security. Security will never be absolutely solved but will be only managed. Instead of defending the entire network as shown from the old defense fashion, the new defense way will be selective and dynamic fashion. For new defense fashion, the end users will also be part of the solution. Moreover, it will be proactive not only defending against the attack from the past as the old fashion. Consequently, the current security aims to control data or network access, prevent intrusion, respond to incidence, ensure network availability, and protect information during the transmission. The security nowadays should concern not only the defenders but also the users themselves.

As mentioned before, security is generally about regulating access to assets such as information or functionality. There is normally a trade-off or conflict between security and functionality or convenience. In addition, security achievement is hard to evaluate when nothing bad happens. The security issue is concerned when the users or the owners want to maximize the availability of their assets. The attackers want to abuse those assets. Therefore, the users or the owners want to minimize the risk by using any countermeasures to reduce these risks. The countermeasures can be non-technical related issues such as physical security of the building, screening of personnel, legal framework to determine criminals, employee training, etc. However, the countermeasures may have vulnerability also causing the risks. The vulnerability is the weakness in security procedures, network design, or implementation that can violate the security policy. The attackers try to have the threats aiming to abuse the assets. The threats can be any circumstance or event with the potential to be harmful to the system. They may exploit the vulnerability to increase the risk at the same time. For this chapter, the risk means the possibility that the particular vulnerability will be exploited. Consequently, the security involves at least four main components including the stakeholders, their assets, the threats, and the attackers. The stakeholders can be the owners, the users, the companies, etc. The assets can be data, services, customer information, personal information, etc. The threats can be erasing, stealing, copying, modifying, etc. Finally, the attackers can be anybody such as employees, clients, criminals, etc.

Although security is about imposing countermeasures to reduce the risks for the assets into acceptable levels, the perfect security is not necessarily costly. As mentioned before, the security cannot be solved, but it can be managed. Consequently, the managing cost depends on a security policy. The security policy is a specification of what security requirements and the countermeasures are intended to achieve. At the same time, security mechanisms to enforce the policy are needed. The necessary actions required to deal with an attack are pre-defined by the security policy. The example objectives of security are such as the confidentiality, the integrity, the availability, non-repudiation for accountability, the privacy, etc. The confidentiality or secrecy means that the unauthorized users cannot read information. The integrity means that the unauthorized users cannot alter or edit information. The availability means that the authorized users can always access information. The non-repudiation for accountability means that the authorized users cannot deny actions. Finally, the privacy means the desire of a person to control the disclosure of personal information. The basic need for security requirement is CIA including confidentiality (C), integrity (I), and availability (A). Moreover, the different security requirements are used for the particular threats. For example, confidentiality is required to protect information disclosure by the unauthorized users. Integrity is necessary to protect changing of information without enough knowledge. Availability is required for Denial of Service (DoS). Authentication is required for spoofing. Access control is needed for unauthorized access. More detail of some important threats is explained in the next section.

5.2 Common Security Attacks and Countermeasures

This section introduces some common security attacks particular for TCP/IP protocol together with some associated countermeasures to provide the basic understanding of security attacks and their importance. TCP/IP protocol is widely used for context-aware applications because it is used to connect the Internet which is the network of the networks.

5.2.1 Security Vulnerabilities

This section introduces the vulnerabilities generally happen in TCP/IP protocol because many context-aware applications require internet connection. The TCP/IP is the two-level package of protocols for the Internet. Since the networks can be completely different such as Ethernet, Asynchronous Transfer Mode (ATM), modem, etc., TCP/IP is designed to connect all those networks together. For the Internet, the routers means the devices from multiple networks. TCP/IP is designed to connect the routers without central control and sophisticated detail. While International Organization for Standard/Open System Interconnection (ISO/OSI) network model has seven protocol layers, TCP/IP has only four layers. ISO/OSI and TCP/IP stack protocol are shown in Fig. 5.1. ISO/OSI consists of physical layer, data link layer, network layer, transport layer, session layer, presentation layer, and application layer. TCP/IP includes link layer, network layer, transport layer, and application layer. For TCP/IP, link layer includes device driver and network interface card. Network layer is responsible for the movement of packets such as routing. Transport layer delivers a reliable flow of data. Application layer provides the details of the particular application. Packet encapsulation is simply done in which the data is sent down through the protocol stack. The heading data is added to each layer.

TCP stands for Transmission Control Protocol sequencing the series of packets to transmit data reliably over the Internet by running on top of IP. IP refers to the Internet Protocol which is the routing of information from the source to the destination. IP is responsible for end to end transmission and sends data in the individual packets. The maximum size of the packet can be varied depending on the networks. If the packet is too large, it can be fragmented. It is unreliable because the packets might be lost, corrupted, duplicated, delivered out of order, etc. TCP/IP is designed for trusted connectivity. Consequently, the attacks can happen on different

Fig. 5.1 ISO/OSI and TCP/TP stack protocols

ISO/OSI Model	TCP/IP Model
Application Layer	Application Layer
Presentation Layer	
Session Layer	
Transport Layer	Transport Layer
Network Layer	Network Layer
Link Layer	Link Layer
Physical Layer	

layers such as IP Attacks, The Internet Control Message Protocol (ICMP) Attacks, Routing Attacks, TCP Attacks, and Application Layer Attacks. Some detail of each attack is introduced briefly in this section.

5.2.1.1 IP Attacks

IP address represents the location in logical network and the identity in the logical host. It is in the form of dotted decimal. The originating host fills in the IP address that will be used for authentication. All devices need to know the IP address that are on the attached networks. The device will send the message directly if the destination is on a local network, the routing is required otherwise. For most of the non-router devices, they just send the data to the local router which knows the network corresponding to each IP address.

IP spoofing is one of the major IP attacks. It is a technique for obtaining unauthorized access to the computers. The intruder sends the messages out with an IP address to make the destination believe that the message is from the trustable host. In the IP protocol, it carries the source IP address and contains port and sequencing information. Every IP packet is routed separately because the IP routing is hop by hop. The route of IP packet is decided by the routers where the packet should go through. IP address spoofing can easily occur because the routers only inspect the destination IP address excluding the source IP address for making the routing decisions. Moreover, the invalid source IP address will not affect the delivery of packets. However, this address will be needed by the destination for responding back. Currently, there are many ways of preventing the spoofing attacks such as avoiding the use of the source address authentication, implementing cryptographic authentication system, configuring the network to reject packets from a local address, enabling encryption sessions at the router when the external connections from trustable hosts are allowed, etc.

5.2.1.2 ICMP Attacks

For IP protocol, there are no built-in processes to ensure that the data is delivered without any problems in network communication. The data is just not transmitted if the router fails or the destination devices are not connected. Additionally, the IP protocol cannot notify the sender if the failed transmission happens. The Internet Control Message Protocol or ICMP is thus the additional component of the TCP/IP protocol stack for addressing this limitation of IP protocol. However, ICMP does not overcome all unreliability issues in IP protocol. Higher reliability can be obtained by using other upper layer protocols. Sometimes, ICMP is just an error reporting protocol for IP protocol. Therefore, ICMP only sends the error report back to the source but not correcting or solving any encountered network problem. Since ICMP messages and any data using IP protocol are encapsulated into datagrams in the same way, more generated error reports can cause more congestion problem. For this reason, the datagram delivery error will never be reported back to the sender of the data. The ICMP protocol can test the availability of the destination by using the echo request message. If the ICMP echo request is received at the destination devices, there will be an echo reply message responding to the source of the echo request. The echo request message is typically initiated using the ping command. In conclusion, the vulnerability can easily happen since there is no authentication for ICMP. ICMP redirects message which can cause the host to the switch gateways. For this reason, this may cause the man in the middle attack and the sniffing.

5.2.1.3 Routing Attacks

The router is used to route the messages when the source and the destination are not on the same local network. Routing is thus based on network address. New route information is required to update the forwarding table. More specifically, this table consists of all information need for routing such as destination, next hop, network interface, etc. A router can be attacked by an attacker in many ways. The commonly found routing attacks (Waichal and Meshram 2013) are distributed Denial of Service attack, Man in the Middle attack, TCP reset attack, etc. This section focuses on introducing Denial of Service attack and Man in the Middle attack.

Denial of Service attack works quickly by making the victim deny of requested service. There are two similar attacks including Denial of Service attack (DoS) and Distributed Denial of Service attack (DDoS). While the DoS is the attack caused by the host on a network, the DDoS is the attack caused by a group of people over the networks. DDoS can be prevented with two different processes including scanning and using the tools. Scanning is to determine the vulnerable hosts that the attack can be carried out. The vulnerable systems can be anyone having no detecting mechanism such as antivirus running or anyone that has not-up-to-date antivirus. After installing these tools on the discovered vulnerable systems, these vulnerable hosts also look for other vulnerable systems for installing the tool on them. This propagation is very quick to cause the DDoS attack on a victim. Some of the DDoS attacks can occur on the router such as Address Resolution Protocol (ARP) poisoning, Ping of Death, and Smurf attack. For ARP poisoning, the ARP request packet in the network is continuously monitored by the attackers. Once it is found, the packet is quickly formed with wrong Media Access Control (MAC) address and sent as a reply. Then, an incorrect mapping will take place in ARP table called ARP poisoning. Therefore, the real MAC will be denied of any further service. For Ping of Death, the attacker forms a special packet which cannot be handled by IP protocol. Therefore, when such a packet is received, it leads to undesirable effects on the victim's machine. For Smurf attack, a lot of ICMP echo request packets are sent out by the attacker to different hosts causing the victim gets flooded with ICMP echo reply packets. Therefore, the victim cannot process any important task because it is busy in processing the echo reply messages.

Man in the Middle Attack (MITM) is the attack that the attacker can intercept the data flowing between a source and a destination. This data can be read or even be modified. More specifically, the MITM attack covers eavesdropping techniques in which the attacker tries to intercept, read, or even alter information transmitting between two or among more computers. On the other hand, these attacks allow the 3rd party to interject themselves and observe the data while passing information back and forth anonymously between the systems. There are several methods for performing the MITM attack. These methods rely on the ability to fool a system to believe that the communication on the network is secured. During normal operations, the computers authenticated by the router allow them to connect to the network, intranet or the Internet. The 3rd party steps in between the destination computer and the router for the attempt to initiate a connection. The MITM then intercepts in communications between the computers. At this position, the MITM acts as a proxy which can read, alter, and insert the data. At the same time, this position allows the MITM to capture transmitted files, public keys, cookies, and passwords passed within the systems. MITM attacks can happen in many situations such as when an attacker is a part of the router along the common point of traffic communications, when the attacker locates on the same broadcast domain as the target, or when the attacker locates on the same broadcast domain as any routing devices which are used by the target. MITM attacks have the potential to interfere communications and confidential information. For this form of attack, any loss of information may be not detectable as the data is read while transferring between systems. Current countermeasures to prevent MITM attacks are, for example, using hardwired connections whenever possible, using the Ethernet cables, utilizing the Virtual Private Network (VPN) connections which operate through Hypertext Transfer Protocol Secure (HTTPS), etc. However, both VPN and HTTPS may not be sufficient enough to secure information.

5.2.1.4 TCP Attacks

For TCP/TP, the sender breaks data into packets and attaches sequence numbers. Then, the receiver acknowledges the receipt and reassembles the packets in the correct order. Then, the lost packets will be sent again. The TCP connections have associated states by starting sequence numbers and port numbers. The sequence number is used for authenticating packets and has a couple of roles. Firstly, if the Syncronization (SYN) flag is set, then this is the initial sequence number. Secondly, if the SYN flag is clear, then this is the accumulated sequence number of the first data byte for the current session of this packet. When the Acknowledgement (ACK) flag is set, the next sequence number which the receiver is expecting is called acknowledgment number. There are three steps for handshaking in TCP/TP. For step 1, the client host sends TCP Synchronization (SYN) segment to the server

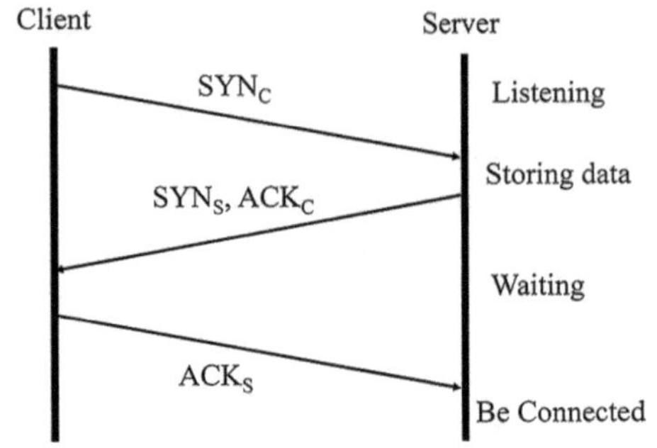

Fig. 5.2 TCP/TP hand checking

and specifies initial sequence numbers without sending any data. For step 2, the server host receives SYN then replies with SYN, ACK segment. Then, the server allocates buffers and specifies the server initial sequence number. For step 3, the client receives SYN, ACK then replies with ACK segment which can contain data. Figure 5.2 shows TCP/IP hand checking steps.

For TCP/IP, the problem occurs when the attackers can learn all important values. Like any MITM attack, if the attacker can learn the associated TCP state for the connection, then the connection can be hijacked. For this case, the recipient may believe that the TCP stream having malicious data, which is inserted by the attackers, comes from the original source. The conceptual diagram of MITM attacks in TCP/TP is shown in Fig. 5.3.

From Fig. 5.3a, the sender and the receiver have established the TCP/TP connection. The attacker who is in the middle between the sender and the receiver can intercept all of their packets as shown in Fig. 5.3b. The attacker can drop all packets from the senders aiming to send to the receiver. At the same time, the attacker can send his/her malicious packets to the receiver instead as shown in Fig. 5.3c. For

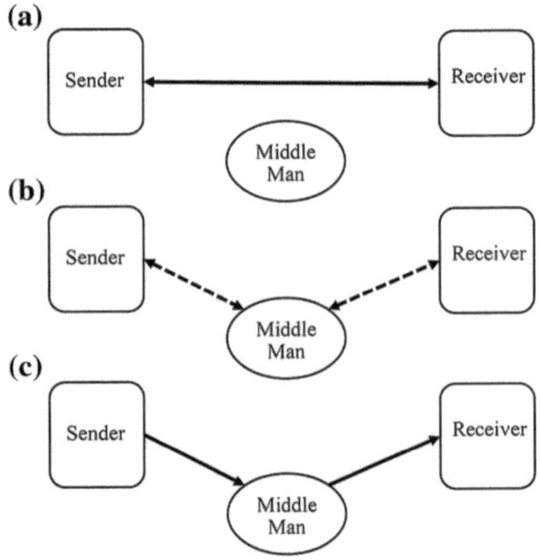

Fig. 5.3 Man in the middle for TCP attacks

example, instead of downloading and running the new program, the users may download a virus. The source authentication and data encryption are required for preventing the system from TCP attacks. Additionally, TCP sequence prediction attack is also crucial because it can be used to identify the packets in a TCP connection and then reassemble the packets. If the attackers know initial sequence number and amount of sent out traffic, they can estimate the current values. Moreover, the classic DoS attack as SYN Flooding is also one of the general attacks for TCP/IP. SYN Flooding happens when the attacker sends many connection requests with spoofed source addresses. Then, the victim allocates resources for all requests. Once the resources are exhausted, the requests from legitimate clients are finally denied.

5.2.1.5 Application Layer Attacks

The common attacks of Application layers are almost the same as of those of the other layers. For example, MITM attack intercepts messages between logical devices, hijacking and spoofing set up a fake device and trick others to send messages to it, sniffing captures packet as they travel through the network, etc.

For application layer attack, there is a particular type of attack by only asking the bots to send requests to the victim for the large files. Then the victim server has to send large files so the bandwidth is saturated and no more requests can be satisfied. This problem can probably be solved if the victim server knows whether the bot or the human is sending the request messages. One popular way is to distinguish the person from the bots by using Completely Automated Turing test to tell Computers and Humans Apart (CAPTCHA). The goal of CAPTCHA is to create the test that is easy for a person to accomplish but difficult for the bots or the computer. Taking advantage of the fact that the humans are good at pattern recognition but the computer is not, CAPTCHAs are expected to distinguish between human and the bots. Originally, CAPTCHAs are the images of distorted text. Since the speech recognition is also difficult for computers to recognize, the new form of CAPTCHAs also provides an audio test for human verification. There are many different alternative forms of CAPTCHAs nowadays, such as text with and without the sound option, picture identification, simple Mathematic CAPTCHA, 3D CAPTCHA, etc. Currently, there are many applications employing CAPTCHAs, for examples, protecting website registration, protecting online polls, preventing comment spam on blogs, preventing worms and spam, searching engine bots, preventing dictionary attacks, etc. However, there are also the vulnerabilities for CAPTCHA. For example, image processing techniques cannot read the text if it is much distorted.

Application layer attacks have the same goal as the other attacks which is to take down the site. These attacks generally include attacking the web server, running PHP scripts and requesting the database for web page loading. Some others application layer attacks are, for examples, cross-site scripting, SQL injection, HTTP Floods, etc. Cross-site scripting enables the attackers to inject the scripts into the web pages which are viewed by other users. SQL injection injects malicious code or SQL query directly into the stringsh, and it will be executed when it is passed to the SQL server. Moreover, the popular DDoS in the application layer is HTTP Floods because one HTTP request can cause the server to run a large number of requests and load various files to create the page. HTTP Floods can be found in many categories. For example, basic HTTP Floods, which are the common and straightforward attacks, try to access the same page over and over again. The randomized HTTP Flood is more sophisticated attack by using the randomization of the Uniform Resource Locators (URLs). The popular victim hosts are gaming, general forum, news, and e-commerce.

5.2.2 *Countermeasures*

This section introduces some countermeasures which are commonly used to protect security for the system.

5.2.2.1 Firewalls

Because there are many types of vulnerability on hosts in the network and the systems cannot be up to date real time, it had better limit the access to the network. As a result, a Firewall is thus chosen for protecting the network. A firewall shares the common idea of the castle having a drawbridge. It is designed to encounter many threats that trying to find a way into the network. Firewall inspects traffic through it and allows only the traffic specified in the policy. Anything outside the policy will be dropped. The conceptual diagram of a firewall is shown in Fig. 5.4.

There are many types of firewalls such as basic router security, packet filters, stateful inspection, and Application Level Gateways or proxy, etc. Primary router security includes Access control Lists (ACLs) and Network Address Translation (NAT). Packet filtering firewall includes packets inspection based on different types of information such as the addresses of the header, the source, and the destination,

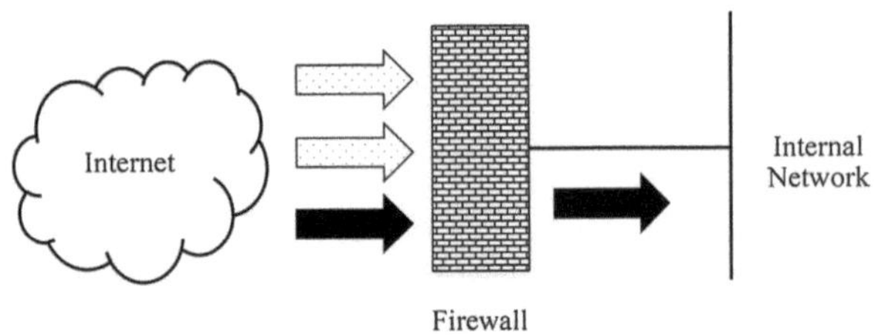

Fig. 5.4 Firewall conceptual diagram

etc. The stateful inspection firewall includes the inspection of the packet based on session information and personal connections. The packets are allowed to pass when it is associated with a valid session and initiated within the network. Application Level Gateway or Proxy firewalls restrict some features and commands from outside the network to protect specific network services. Additionally, many operating systems also have the built-in firewall. Some routers even come with firewall functionality. Moreover, firewall can be hardware or software.

Among those firewalls, the packet filter firewall is commonly used with the simple assumption. More specifically, it selectively passes packets from one network interface to another. It is usually done within a router between the external and the internal network. The filter is based on both packet header field and packet content. The header fields are, for examples, IP source and destination addresses, application port numbers, ICMP message types or protocols, etc. The packet contents are the payloads. The possible actions are to allow the packet to go through, drop out the packet, alter the packet, and log information about the packet. The rules based on condition and associated actions are used for implementing firewalls. The packet filter firewall is transparent to the applications and the users. However, some benefits are the issues of security, speed, and usability. For proxy firewall, the available data is application level information such as user information. There are some advantages such as better policy enforcement and better logging. However some disadvantages are, for examples, this type of firewall does not always perform well, one proxy is used for each application, and it is easy for modification, etc.

Because there is only one point of access to the network, the firewall can be positive and negative at the same time. It is positive because any circumstances coming in and going out can be traced and recorded. Another way around, it can be negative that the system can be cut off from everyone else if this point is broken. This single point may also cause the problem of congestion. Although firewalls can ensure the security for some levels, they may also cause the underlying problem because it is difficult for the users to keep up with changes and always keep host secure.

5.2.2.2 Intrusion Detection System

Intrusion is the attempt to break into the system. The intruders can be anyone from the outside the network or the legitimate users of the network. Moreover, the intrusion can be a physical, system, or remote intrusion. Intrusion Detection System (IDS) looks for the signature of the attack. The attack signature is usually the distinct pattern indicating any malicious or suspicious intention. The attack signatures are, for examples, ping sweeps, port scanning, web server indexing, OS fingerprinting, DoS, attempts, etc. Another speaking, the IDS can protect the network against the known software. It is also useful for monitoring of suspicious activity on the network. However, the intrusion detection is only useful if the attacks are occurring as planned.

The IDS monitors the operation of all key components in the network such as firewalls, routers, key management servers, and files, etc. It allows the administrator to tune, organize, audit, and logs into the trails quickly. It can help non-technical staff to perform the security management of systems by providing the user-friendly interface and the extensive attack signature database. It can also recognize and report alterations to the data files. Because the firewall cannot detect security associated with traffic that does not pass through it, the system only with firewalls might not be safe due to many reasons. For example, not all accesses to the Internet occur through the firewall. The firewall does not inspect the content of the permitted traffic. The firewall is usually helpless against tunneling attacks. For this reason, the IDS is required to increase the awareness of traffic on the internal network. In summary, IDS is additional need for the system because it is capable of monitoring messages from other pieces of security infrastructure.

There are many different ways of classifying an IDS such as anomaly detection, signature based misuse, host-based IDSs, and network-based IDSs. For anomaly based IDS, it models the typical usage of the network as a noise characterization. Therefore if anything differs from the noise, it will be assumed to be the intrusion activity. The primary strength of this IDS is its ability to recognize the novel attacks. However, this type of IDS works well based on the assumption that the intrusions are sufficiently unusual so they can be clearly detected. This limitation can generate many false alarms and affects the effectiveness of the IDS. Next, the signature based IDS has the attacked description that will be used to match with the sensed attack manifestations. However, this type of IDS cannot detect novel attacks regardless their descriptions. It also frequently suffers from false alarms. It needs to be programmed again and again for every new pattern. Next, the host-based IDS logs in the audit information and analyze this information to detect the trails of the intruder. This audit information can be the usage of identification and authentication mechanisms, file opening and program executions, administration activities, etc. However, logging in this kind of information requires high experience. Moreover, selective logging runs the risk that the attack manifestations could be missed. For network-based IDS, it looks for attack signatures in the network traffic. There is a filter applied to determine which traffic will be discarded or passed to the attack recognition component. This filter helps to screen out the known and no malicious traffic. Although IDS seems like the additional requirement of the system besides firewalls, it doesn not mean that the IDS is the complete solution for security system because of many other concerns. For example, IDS requires the human intervention to conduct the investigations of attacks. It cannot intuitively learn the contents of security policy. It cannot compensate all weaknesses in the network protocols. Although it is capable of monitoring network traffic, it cannot go to some traffic level. Therefore, to satisfy the requirements of security, other security countermeasures are carefully selected accordingly to targeted context-aware applications.

5.3 Security Recommendations for Context-Aware Applications

The security required for context-aware systems and applications have been specified briefly by International Telecommunications Union (ITU-T) presented in their recommendation (Almutairi et al. 2012). The examples of concerning security issues are shown in this section.

5.3.1 Access Control

Access control is to give permission to the authentic and real users to access the system facilities. It aims to restrict access to the applications from unreal or unauthorized users. The users have to be verified to satisfy the privileges so that the users can receive authority to access the resources. However, if the users are not satisfied the privileges, the request will be rejected (Bardram et al. 2003). Access control involves authentication and authorization. While authentication is the restrictions on whom or what can access system, authorization is the restrictions on actions of authenticated users. This section briefly describes the methods for both authentication and authorization.

5.3.1.1 Authentication

Authentication is considered as the basic requirement of verifying the identity of the entity for every node in the system. In security systems, there is the difference between authentication and authorization. Authentication is the process of allowing the individuals to access to system objects accordingly to their identity. However, authentication mainly ensures that the person is really who he or she claims to be without checking about the access rights of the individual. For context-aware applications, every node is required to verify the identities of the communicated entities in the network to ensure that the nodes are communicating with the right entity and the users who are attempting to access the applications is reliable or eligible for accessing. Only the authorized user can access the applications. Security infrastructure requires mutual authentication to validate the user's identity.

Authentication Method

Authentication can be done in many ways. According to the distinguishing characteristics, the authentication methods can be classified into three types (Menkus 1988) including knowledge-based authentication which is the authentication from what the user knows, the possession-based authentication which is the authentication from what the user uses, and the biometric-based authentication which is the authentication from what the user is. The examples of the knowledge-based authentication are the password, the personal identification number (PIN), the passcode, etc. The possession-based authentication means the authentication from memory card and smart card tokens. Finally, the biometric-based authentication typically includes physiological characteristics such as face, iris, fingerprint, etc. or behavioral characteristics such as keyboard and mouse dynamics, etc. Although all types of authentication methods are usually used in any context-aware application, the biometric-based authentication is gaining more attention for context-aware applications nowadays because it can satisfy the requirement of automatically gathering of user context.

At the same time, authentication protocols are capable of authenticating the connecting parties. This kind of protocol is gaining more attention nowadays to provide the secure connection among devices or nodes within the context-aware applications. The most crucial decision of designing secure systems for context-aware applications should be the selection of an appropriate authentication method. For choosing the right authentication method, some aspects have to be taken into account. For example, the developers need to decide the desired level of security. The developers need to also concern about the complexity of the used techniques because it consumes more power, speed, maturity of the technology, scalability of technology. Moreover, the developers should also concern about the practicality of the used methods such as they do not cumbersome updating and are user-friendly. This section thus overviews some authentication methods and authentication protocols used in general security system and context aware applications.

Knowledge Based Authentication

This type of authentication is simply to use. However, there are also many concerns about its efficiency.

Passwords

Passwords is the most widely used form of knowledge-based authentication. A password is simple for both the system designers and the end users. The users provide an identifier which is a typing in word or phrase frequently called user identification along with a password. The password is encrypted. Password authentication does not typically require complicated method or complicate hardware since this type of authentication is mainly simple and does not consume much processing power. At the same time, password authentication also has several vulnerabilities. For example, an eavesdropper might see the password if it is sent in the clear format. The intruder may have a chance to read the password file on the server. A password may be easy to guess by the attackers after making several attempts to log in. A password may be crackable using encrypted recognizable items from offline guessing attack. At the same time, there are some other concerns about password authentication. For example, the administration might give the tighten rules for generating the password leading the inconvenient use for the users. The passwords must be short enough to remember and encrypted at the same time. The server should disable client's account after too many failure password attempts.

This task can help protecting the password from online attack. However, it is difficult to protect it from off-line attack because the attacker gets some relevant information which can be used to crack the encrypted passwords by repeatedly trying passwords until the agreement with data is reached.

The passwords can cause the limitation of human information processing (Yan et al. 2004; Sasse et al. 2001). Therefore, the passwords should be any set of characters that are difficult to guess and easy to remember at the same time. This requirement is not easy to achieve because the passwords that are difficult to guess are frequently difficult to remember. Moreover, most users have to remember multiple passwords for different systems and applications. Therefore, the users usually choose meaningful strings such as names or nicknames which are easy to remember and definitely to crack (Adams and Sasse 1999) and they usually duplicate their passwords (Ives et al. 2004). To improve password security and protect it from attacks, the password policy should be implemented (Smith 2002). Some major rules from password policy are guided. For example, there are non-dictionary and no-name passwords. The password should be long enough with mixed different types of characters. The complex passwords should use acronyms, rhymes, and mnemonic phrases (Carstens et al. 2004). The passwords should not be shared and should not be given to other people or even written down. The passwords should be encrypted or hashed, etc. Passwords based on the rules as mentioned above are more efficient, harder to identify and to determine. To address the problem of sniffing passwords, one-time passwords are widely used when authentication is performed over the Internet.

One-time Passwords

One-time password can be obtained by a challenge-response password and a password list (Duncan 2001). The challenge-response password provides the response with a challenge value after receiving a user identifier. The response is calculated from either the response value with some electronic devices which is one type of possession-based authentication or from a pre-defined table. On the other hand, a one-time password list uses lists of passwords which are sequentially utilized by the users. The values are generated and cannot be calculated easily from the previously presented values.

Possession Based Authentication

Possession-based authentication can be considered as the authentication method based on what the user has. The physical objects called tokens are commonly used for this kind of authentication. An obvious problem is the unconvinced usages for the users because they have to carry the token all the time they need. There is also the risk of being stolen. Tokens are usually found into different types including memory tokens and smart tokens. The memory token is cheap and easy to use. It just keeps the information without any data processing. It is used together with either passwords or PIN which is a knowledge-based authentication mechanism to provide more security than using passwords or PINs alone. On the other hand, the smart tokens can perform data processing. They are widely used because of their portability and cryptographic capacity (Juang 2004; Kumar 2004). Although they are more secure than memory one, they are usually more expensive and much harder to use.

Biometric Based Authentication

Biometric is the scientific discipline of measuring relevant attributes or characteristics of living individuals to identify properties or unique features of each individual. It can be used for security because those unique features can distinguish one person from another and that theoretically can be used for identification or verification of identity. Biometric-based authentication is the method authenticating the user based on what the user is. It is used for automatic identification by using anatomical, physiological or behavioral features, and characteristics associated with the users (Kim 1995; Wayman et al. 2005). The emergence of biometric authentication helps to obtain more efficient and accurate identification because it cannot be easily stolen or shared. Moreover, the biometric based authentication uses the unique characteristics of the person to perform the identification. However, biometrical based authentication is usually expensive because of the technical complexity and the requirements of specialized hardware or sensor. There are also the concerns about ethical issues of potential misuse of personal biometrics. Therefore, it is normally used for the applications with high levels of security protection such as tracking, surveillance, etc.

Biometric applications available today are categorized into two key features including physiological and behavioral characteristics. Physiological biometrics is based on the user's physical attributes which are usually stable. The well-known of physical characteristics are fingerprints, finger scans, hand geometry, iris scans, retina scans, facial scans, etc. Fingerprints are the most widely used for all among physiological characteristics (Snelick et al. 2005; Tuyls et al. 2005). However, there are some disadvantages for fingerprints. The dirt, grime, and wounds together with the placement of fingers may cause the error detection. The fingerprint detection typically requires the vast database to process. For hand geometry recognition, it uses the geometry of users' hands for recognition. It is usually more reliable than fingerprint recognition. However, it requires an enormous scanner which is not convenient for all applications. Retinal scanning is also a favorite application. The user looks straight into retinal reader scanning with low-intensity light. It is very efficient and cannot be spoofed. However, the user has to look directly into the retinal reader to avoid the error detection. Iris scanning also no touch required method. It scans unique pattern of iris which is colored and visible. However, contact lenses can be an issue of accuracy and inconvenient usages. For face recognition or scanning, the user needs to face the camera and the neutral facial expression is required. The error detection frequently depends on lighting condition and facial position. It is also easy to be spoofed. Moreover, it also requires big data storage and complex algorithm for data processing and dealing with identification across facial expressions.

On the other hand, the behavioral biometrics is based on behavioral attributes of the users that are dynamic attributes instead of static attributes as physical characteristics (Guven and Sogukpinar 2003; Saevanee and Bhatarakosol 2008; Frank et al. 2013). The well-known methods are speech or voice recognition, signature recognition, typing and touching pattern recognition, etc. Speech or voice recognition has the user's natural tone speech as the input. It can be considered as the most user-friendly method. The disadvantage is that this method requires the sophisticated and robust algorithm for dealing with illness and emotional behavior. Additional algorithms for eliminating background noise and deal with device quality are needed to improve the efficiency. The signature can measure the dynamic patterns of speed, velocity and pressure pattern of the users. It is one of the most accepted methods for the users. Unfortunately, the signature is variable with many factors such as age, illness, emotions, etc. At the same time, the user's typing or touching pattern such as speed, pressing and releasing rates can be measured and kept to generate the personal and unique patterns. Typing and touching pattern recognition is not very scalable. Additionally, it can be spoofed by the simple technology of the recorders.

The biometric authentication processes consist of two primary processes including the collection of master characteristics and the verification of those master characteristics. More accurately, the collection of master characteristics involves the biometric acquisition, the creation of master characteristics and storage of master characteristics. For verification, it requires biometric acquisition, comparison, and decision-making. The biometric acquisition involves the sensors. The personal attributes of the users are captured and stored for the next authentication. The accuracy of each biometric system can be done by many measurements of biometrics including erroneous rejection or false non-match (type I error), and incorrect acceptance or false match (type II error). For any biometric application, both error types are subjected to be as low as possible. Current applications of biometrics authentication can be found in many application domains especially banking and immigration facilities. There are also some concerns regarding the biometric characteristics that they are not encrypted and depends heavily on input devices. Also, they cannot authenticate computers or any smart devices.

5.3.1.2 Authorization

Authorization is a form of access control for action restrictions of the authenticated users. Access control matrix is the simplest form to do authorization. This matrix has all relevant information such as all users and all resources, etc. The user is checked with this matrix before access to any allowed resource. Authorization is enforced by access control lists and capability lists. The example of access control list is shown in Fig. 5.5. The case of capability list is presented in Fig. 5.6. It can be seen that Access Control Lists (ACLs) stores access control matrix by column while Capabilities (C-Lists) stores access control matrix by row. From Fig. 5.5, it can be seen that User1 is only one who can read, write and delete the Payroll data. From

Fig. 5.5 Example of matrix with access control list

	Mortgage Data	*Payroll Data*	Credit Data	Insurance Data
User1	rwx	*rwx*	rw	rw
User2	rx	*rx*	r	rw
User3	rx	*rx*	rw	rw
User4	rx	*rx*	r	r

r: read
w: write
x: delete

Fig. 5.6 Example of matrix with capacity list

	Mortgage Data	Payroll Data	Credit Data	Insurance Data
User1	rwx	rwx	rw	rw
User2	rx	rx	r	rw
User3	rx	rx	rw	rw
User4	*rx*	*rx*	*r*	*r*

r: read
w: write
x: delete

Fig. 5.6, it can be seen that User4 can read and delete Mortgage and Payroll data while he/she can only read Credit and Insurance data. The protection is data-oriented and easy to change rights to a resource. The C-Lists are easy to delegate, add, or delete the users. It is simpler to avoid the confused deputy. However, it is harder to implement.

5.3.1.3 Multilevel Security Models

Multilevel Security Models (MLS) is a form of access control and has been widely used in many application domains such as government, military, business, healthcare, cyber security, etc. For example, the information in the bank can be restricted differently to executive board, all management departments, everyone in the bank, and general public. For cyber security, the intruders are assigned at the low level to limit the damage. As firewall must decide what can be in and out, it is one form of access control. More specifically, MLS is the capability of a computer system that can allow the users to simultaneous access with various security clearances. It also prevents the users from obtaining access to informationwithout authorization. The sensitivities can be classified information at different security levels. Additionally, it is typically used for mandatory access control for achieving primary security as confidentiality. The well-known example of security levels is the National Security Agency (NSA) security manual that classifies data into Top

Secret, Secret, Confidential, and Unclassified security level. In this case, it can be seen that Top Secret level is more secure than Secret level. The Secret level is more secure than Confidential level. The Confidential level is more secure than Unclassified level. The confidentiality ensures that the information does not flow to those not cleared to that level.

In general, three main models can be used for access control such as Discretionary Access Control (DAC), Mandatory Access Control (MAC), and Role-Based Access Control (RBAC) (Almutairi et al. 2012). For DAC (Harris 2010), the system with DAC allows the owner of the resource to specify which subjects can access which resources. As a result, this model is called discretionary because the access control is based on the owner's discretion. DAC systems use the identity of the subject, which can be a user identity or group membership, to allow or to deny access. The easy way of DAC implementation is through ACLs, which are set by the owners. Additionally, the DACs can be used for both the directory tree structure and the files it contains.

For MAC (Harris 2010), the users and the data owners do not have freedom to determine who can access which resources. Instead, the operating systems enforce the security policy of the system through the use of security labels. This type of model is used in the applications that both information classification and confidentiality are very critical as a military installation. The final decision is made by the operating system and it can even override the expectations of the users. MAC uses a security label system. The users have clearances, while the resources have security labels containing data classifications. MAC uses these attributes to determine access control capabilities. The users are given a security clearance, and data is classified with the same way. The users have access to data classified as equal and less than their status. Every subject such as the file, the directory, and the device has its security label together with its classification information. For the first use of MAC model, all subjects and objects must have sensitivity labels containing classification categories. The classification indicates the security label, while the categories identify need-to-know rules. The well-known MAC Model is Bell-LaPadula Model (Bell 1996) enabling one to formally show that a computer system can securely process classified information.

For RBAC (Harris 2010), it is also called nondiscretionary access control. It uses a central set of controls to determine how subjects and objects interact. This type of model uses the role of the user within the organization to allow the access to the resources. General speaking, the RBAC approach assigns the access control by allowing the permissions based on roles of the user. The role is commonly defined regarding the operations and tasks. It can be seen that introducing roles also implies the difference between rights assignment for both explicitly and implicitly. Explicit assignment indicates that the roles are assigned to the particular individual. On the other hand, implicit assignment indicates that the roles are assigned to a role or group, and the user inherits those attributes.

5.3.2 *Privacy and Confidentiality*

These requirements is to protect or restrict the use of some high sensitive, private or secure information from being shared or being available to anyone else without the permission from the owners. While the privacy is about people, the confidentiality is about data. Privacy means preventing the identity from being disclosed to any other entities. The confidentiality means keeping the data from being revealed to the entities that have no permission to access it. More specifically, the privacy is about people and their sense of being in control. Confidentiality is a treatment of private information. It is usually based on the belief that it will not be revealed except there is the agreement previously done. Maintaining privacy and confidentiality helps to protect the users from different kinds of harms such as embarrassment, social harms such as unemployment, financial credit, and criminal liability, etc. For the context-aware application, there is the requirement to have both privacy and confidentiality when the data is exchanged among all computing nodes. Implementing this requirement is very challenging because the user normally wants to access all functionalities and facilities of context-aware applications, but they would not always want to share their contextual information. Therefore, some methods to protect the confidential information from unwanted users are required.

5.3.3 *Data Integrity*

Data integrity is to ensure that the data being sent through context-aware applications should be received by the intended entities without being changed by unauthorized modification (Almutairi et al. 2012). This requirement is essential for especially in military, banking, healthcare and aircraft control systems because the modification of data would cause enormous damage. For context-aware application, the integrity requirement is defined as the access to the user's resources is not allocated or assigned to any illegal or incorrect user. In cloud storage which is commonly used by any current context aware application, the data integrity is very crucial. Cloud storage keeps the user's data to the massive data centers. The users can access their data remotely anywhere and anytime. It is always available and highly mobile and available across platforms. Cloud reduces the cost of deployment and the risk of data loss. Clouds also require fewer maintenance concerns because the software and the hardware do not require installing or upgrading very often. However, cloud storage can cause many new security challenges especially for promoting the collaborative workspace in real time. More sophisticated data scheme is indeed required such as the proof of irretrievability (POR) using cryptographic algorithms (Devika and Jawahar 2015). This section introduces some of the basic techniques of the cryptographic data algorithms including Data Encryption Standard (DES), Advanced Encryption Standard (AES), Rivest-Shamir-Adleman (RSA) algorithm, the Secure Hash Algorithm (SHA), and Hash (HASH) Algorithm.

DES (Stallings 2006) is a symmetric block cipher. This algorithm uses a 56 bit key to encipher or decipher a 64 bit block of data. The key is presented as a 64 bit block. Every 8th bit of the key is ignored. In 2001, AES was firstly introduced. It is also a symmetric block cipher, and it is intended to replace DES be-cause of the standard is suitable for many applications. The AES cipher forms the latest generation of block ciphers. Its block size has increased from 64bits to 128 bits. The keys have increased from 128 to 256 bits. The DES is claimed as the secure method, but it is very slow. At the same time, there are various key lengths provided. By using AES, the amount of control on encrypted data depends only on the type of encryption. Some particular encryption modes can disturb the attackers to perform modifications. Next, RSA is a public key algorithm (Rivest et al. 1978) performing encryption and decryption by using different keys. Therefore it is called as the asymmetric block cipher. The RSA algorithm employs the public key for encryption, and the private key to decryption in the digital signature technique. More specifically, the RSA is very slow because of longer keys comparing to a symmetric block cipher as DES. For SHA, it is the most widely used hash function. Finally, HASH is used for computing a condensed representation of a fixed length message. This message is sometimes known as a message digest or a fingerprint. It can cause the excessive collisions which lead to the poor performance.

5.4 Security Protocol

While human protocols are the rules for the human to follow in their interaction, the networking protocols are the rules to follow in networked communication systems. The examples are such as HTTP, FTP, etc. The security protocol is the communication rules followed in a security application such as Secure Sockets Layer (SSL), Internet Protocol security (IPSec), Kerberos, etc. The ideal security protocol is to satisfy security requirements. It has to be efficient use such as minimizing of computational demand, costly public key operations, delays or bandwidth, etc. It must work when the attackers try to break it and even if the environment changes. Also, it needs to be easy to use and implement. However, it is difficult to satisfy all requirements practically. The simple security protocols are; for example, secure entry to the room and ATM machine protocol. An authentication protocol is a message sequence exchanging between entities that either distributes secrets or allows the use of some secret (Burrows et al. 1989). Authentication on a stand-alone computer is relatively straightforward because the primary concern is an attack on authentication software. On the other hand, performing authentication over the network is much more complex. The attacker can passively observe messages, re-play messages and actively attacks by inserting, deleting, or even changing the messages. This kind of attack is called replay attack in which the conceptual diagram is shown in Fig. 5.7. It can be seen from Fig. 5.7a that the sender wants to communicate with the receiver. The receiver asks for the password to identify the sender. The attacker can watch the sender's password while

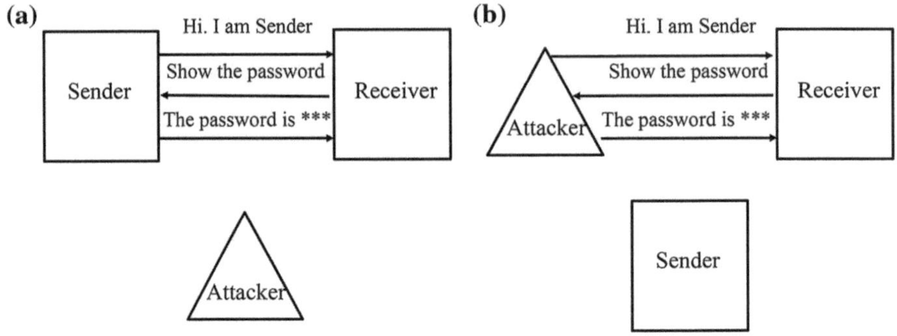

Fig. 5.7 Conceptual Diagram of Replay Attack

communicating. Then for Fig. 5.7b, the attacker communicates with the receiver with the sender's password. To prevent replay, the challenge-response is used such as something only the sender and the receiver know, etc.

In this section, some of the common used protocols for addressing security issues within open networks are introduced. The protocols introduced in this section aim to provide the basic understanding that the security can be protected by almost different network layers such as Secure Sockets Layer (SSL), Secure Shell (SHH), etc. TCP-based authentication as IPSec is also explained. Although TCP is not intended for using as an authentication protocol, IP address in TCP connection is often used for authentication. Since SSL is the protocol used for most secure transactions over the Internet, it is worth to study the foundation of internet security. Moreover, SHH is a good example of application layer protocol which is useful for extending to the development of security protection for context-aware applications.

5.4.1 Secure Sockets Layer

Secure Sockets Layer (SSL) becomes an internet standard in 1996 to provide a secure method of communication for TCP connections (Hickman and Elgamal 1995), especially for HTTP connections. More specifically, SSL is a cryptographic protocol to secure network across a connection-oriented layer. Any program using TCP can use SSL connection with some modification. SSL is flexible for selection of symmetric encryption, message digest, and authentication. It is a layered protocol which operates between the Internet TCP protocol and application protocols as shown from the conceptual diagram in Fig. 5.8.

SSL is used to authenticate the server to the client and allows the client and the server to select the suitable cryptographic algorithms or ciphers freely. Optionally, it also authenticates the client to the server. SSL uses public-key encryption

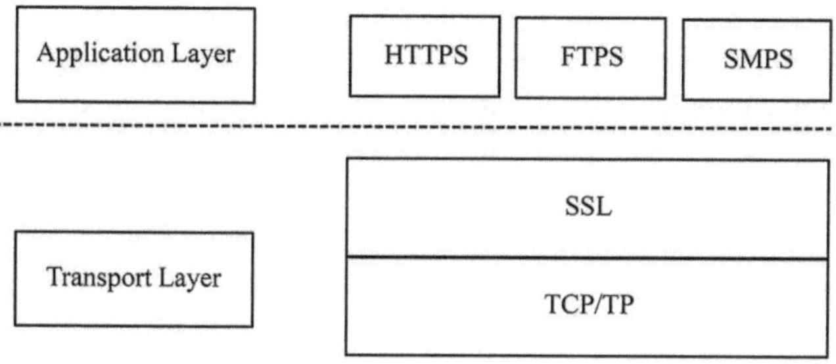

Fig. 5.8 Conceptual Diagram of SSL Protocol

techniques to generate the shared secret and establishes an encrypted SSL connection. Then, the client is allowed to authenticate itself to the server. Finally, it uses the hello message to establish an encrypted connection to both the server and the client. After exchanging the hello messages, the certificate from the server will be sent. A certificate contains certificate issuer's name, the entity to whom the certificate is being issued, the public key of the subject and the time stamps. When the server has been authenticated, the server then requests the certificate from the client. After receiving hello message from the client, the server tells the client to start using encryption to finish the initial handshake. Now, the application can perform transferring. Only the hello messages are exchanged when the client and the server decide to resume the previous session or duplicate the existing session. The advantage of resuming the previous session is to save the processing time which will effect on server performance. SSL protocol architecture is shown in Fig. 5.9.

It can be seen that SSL includes many sub-protocols. Record Protocol defines the format which is used for transmitting the data. More specifically, the compression and the decompression occur in Record protocol. Handshake protocol verifies the server and allows client and server to agree on an encryption set before transmitting the data. Alert protocol is responsible for the severity of the message and a description. Change Cipher Spec protocol notifies the other parties to use the new cipher suite before the finished message. For the SSL protocol, RSA public key cryptography is used for Internet security. A pair of asymmetric keys is used in public key encryption for performing both encryption and decryption. It means that each pair of keys consists of

Fig. 5.9 Conceptual Diagram of SSL layer

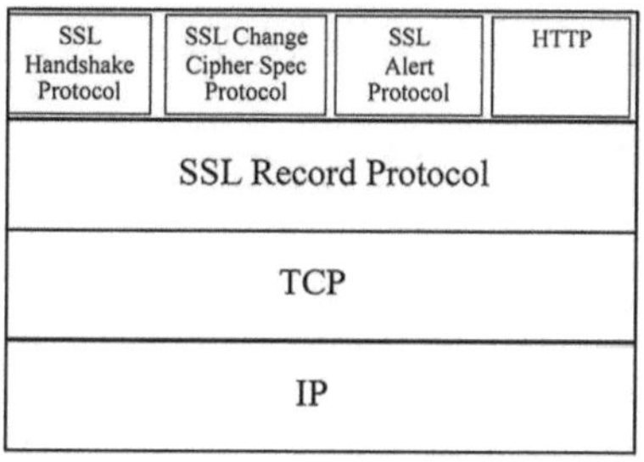

both a public key and a private key. The public key can be distributed widely while the private key has to be stored secret. Generally, data encrypted with the public key can be decrypted only with the private key. At the same time, the data encrypted with the private key can be decrypted with the public key only.

5.4.2 IP Security

IP Security or (IPSec) is the security at the network layer. It provides a set of security algorithms together with a general framework that allows any communication between entities. It can use any algorithms to ensure the safety for the communication. The applications of IPSec are, for example, to secure remote access to the Internet, to establish extranet and intranet connectivity with partners, etc. The benefit of IPSec is that it is transparent to applications which are below transport layer and provides security for individual users. Internet Key Exchange (IKE) is a method for establishing a security association (SA) for authenticating the users, negotiating the encryption method and exchanging the secret key. IKE ensures the secure transmission of the secret key to the recipient by using public key cryptography. It is also used in the IPsec protocol.

More specifically, IPSec is general IP security mechanism consisting of three functional areas (Stallings 2006) including authentication, confidentiality and key management. The authentication mechanism assures that the pre-identified party transmittes a received packet and that packet will not be altered during the transmission. The confidentiality facility enables communicating nodes to encrypt messages to prevent eavesdropping by the third parties. The key management facility ensures that the keys are exchanged safely. It is also applicable to use over many different networks such as LANs, across public and private WANs, the Internet, etc. In a firewall and router, IPSec provides high security to all crossing traffics. It can also ensure the safety even for the individual mobile users. The network identity is hidden with IPsec. It provides basic security requirments including confidentiality, authenticity, and integrity. It can connect sites with more cost-effective and secure network than those with the leased lines. Additionally, it allows the user to work from home and the mobile hosts. However, IPSec can be a single failure point in the path that can disconnect the entire network and cause the bottleneck problem.

5.4.3 Secure Shell

With the evolution of the Internet, services such as file transfers, remote logins, and remote command executions became possible. Some problems have existed with some supported protocols, such as ftp, telnet, etc. because they lack the security. It is highly possible for the intruder to intercept and read data. Particularly, telnet is risky because the plaintext, the username, and the password are easily intercepted over the network. Therefore, there is the necessary to have the protocol to address this problem as Secure Shell (SSH). SSH is a protocol facilitating the secure remote login and other secure network services over the insecure network. It is the security at the application layer. The application can understand the data and can provide the appropriate security by extending application without involving operating system. However, the security mechanisms have to be designed independently of each application.

More specifically, SSH can be considered as both a program and protocol. It allows the users to log into another computer over an insecure network, executes commands and transfers files. It was developed for the replacement of telnet, ftp, and some others. It uses TCP and provides authentication, confidentiality for data and command, integrity, authorization, and data compression. It has transparent communication between the client and the server over encrypted network connections. It can be implemented on almost operating systems. SSH has many important features. For example, it performs authentication by proofing of identity of the users and the servers with the common password and public-key signature. It ensures privacy with robust standard encryption algorithms. For integrity, the cryptographic integrity checking is used. SSH can have multiplex services over the same connection. More importantly, it is free for non-commercial use. In conclusion, the most significant advantage of SSH probably is its protection against packet spoofing, IP/host spoofing, password sniffing, and eavesdropping. However, there are also some disadvantages. SSH only supports known port numbers. Moreover, SSH cannot fix all TCP's problems because TCP runs below SSH. SSH cannot protect the users from being attacked from other protocols.

5.4.4 Wireless Network Security

Wireless networks are rapidly becoming pervasive. It is the way that a computer is connected to a router or is linked to other computers and devices without a physical link. The attacker may hack a victim's personal device and steal private data to perform some illegal activities by using the victim's personal identification. It is also possible that the attackers can read the transferred data by using the sniffers. For wireless security, the concerns are similar to those in a wired environment, such as an Ethernet LAN or WAN. The security requirements are the same such as confidentiality, authenticity, integrity, availability, and accountability. However, some of the security threats are worse than those in the wired network environment, and some are unique in the wireless environment. The risk in wireless networks is the communications medium and also from the traditional protocols. The key factors (Stallings and Brown 2008) contributing to the higher security risk of wireless networks are, for example, channel, mobility, resources, accessibilities, etc. For channel, wireless networking typically involves broadcast communications, which is more vulnerable to active attacks exploiting vulnerabilities in the communications protocols. For mobility, higher risks come from many portable and mobile devices in the network. For resources, many wireless devices have limited memory and processing resources, so they have more risk to encounter threats such as malware and denial of service. For accountability, some wireless devices may be left unattended causing higher vulnerability to the physical attacks.

Some serious wireless network threats (Stallings and Brown 2008) are, for example, accidental association, Ad hoc networks, malicious association, MAC spoofing, MITM attacks, DoS, network injection, etc. For accidental association, the users sometimes unintentionally log into the neighbor wireless access point because of the overlapping transmission ranges. For Ad hoc networks, some peer to peer networks can easier have security threats because there is no central control point. For nontraditional networks, some new networks, such as personal network Bluetooth devices, have highly potential to have security risk regarding both eavesdropping and spoofing. For malicious association, a wireless device appears to be a legitimate access point which can steal passwords from the legitimate users and then use a legitimate wireless access point access the network. For identity theft or MAC spoofing, this can occurs when an attacker eavesdrops the network traffic and can identify the MAC address of a computer. For MITM attacks, this attack persuades a user and an access point to believe that they are communicating to each other. However, the communication is going through the intermediate attacking device instead. For DoS, this attack occurs when a wireless access point is attacked with various protocol messages which are designed to consume system resources. For network injection, this attack aims to degrade network performance by targeting wireless access points that are exposed to non-filtered network traffic.

There are some general recommendations for wireless network security to be employed for any system and application including context-aware applications. For example, the system must use encryption. For router-to-router traffic, the built-in encryption mechanisms are normally attached with the wireless routers. All anti-virus and anti-spyware software have to be up-to-date. The broadcasting identifier should be turned off to avoid thwart attackers know the identity of the routers. The identifier of the router should be changed from the default. The router's pre-set password for administration should also be modified. The router should be configured only communicates with specific computers which are already approved MAC addresses.

Because of some differences between wired and wireless network, the robust security services and mechanisms, particularly for WLANs, are required. The security services and mechanism for WLANs have been evolved along the way the growing of security requirements. The original 802.11 specification includes a set of weak security features for privacy and authentication (Eissa et al. 2013). For privacy, IEEE 802.11 defines the Wired Equivalent Privacy (WEP) algorithm. WEP is a protocol to protect link-level data between clients and access points (Eissa et al. 2013). For ensuring authentication, WEP provides the access control to the network by allowing the access to client stations that passes the authentication correctly. For ensuring confidentiality, WEP prevents information from casual eavesdropping. For ensuring integrity, WEP prevents the messages during the transition between the wireless client and the access point. Later, the Wi-Fi Protected Access (WPA) has been introduced. WPA is a set of security mechanisms that eliminates most 802.11 security issues and is based on the IEEE 802.11i standard (Eissa et al. 2013). Then, the Robust Security Network (RSN) has been later introduced. The Wi-Fi Alliance certifies several vendors in compliance with full IEEE 802.11i specification under the WPA2 program. Choosing the right technology for any system and application is still challenging because there are many involved factors such as data rates, interoperability, etc.

5.4.5 Wireless Sensor Network Security

A sensor network is a heterogeneous system having multiple computing elements working together such as the tiny sensors and the actuators. Most sensor networks consist of a large of low power and cost nodes interacting with the environment. They are widely applied in many applications such as manufactured machinery control, building safety, earthquake monitoring, military applications, medical and healthcare monitoring, ocean and wildlife monitoring, etc. The need for security of WSN is increasing nowadays because it is becoming the practical and cost-effective way for deploying sensor networks. Since it has different challenges as compared to traditional networks, new and different mechanisms are necessarily employed.

Some critical security requirements of WSNs are data confidentiality which is the most important issue in any network, data integrity, data freshness which ensures that no old messages will be replayed, and data availability. For WSNs, there is the need to adjust existing encryption algorithms to fit a WSN for ensuring data availability. This requirement may cause additional computation and communication which consume more energy. A WSN requires every node to be self-organizing and self-healing (Albers et al. 2002). The secure localization, which is the ability to accurately and automatically locate each sensor in the network is crucial for WSNs. Additionally, authentication is also indeed required. WSN is vulnerable to many types of attacks such as DoS, traffic analysis, privacy violation, physical attacks, etc. DoS can jam a node or set of nodes by transmission of a radio signal interfering with being used radio frequencies. More attacks can be found (Kalita and Kar 2009) such as Sybil attack which is a malicious device illegitimately taking on multiple identities, node replication attack, physical attacks, etc.

Although the security requirements of WSNs are to shares some common requirements with the traditional networks, many new requirements emerge because of some limitations and the unique characteristics. For example, sensor nodes are often deployed in open areas allowing the fast physical attack. Since sensor devices are limited in their energy, computation, and communication capabilities, the existing public-key cryptographic is too expensive regarding system overhead (Perrig et al. 2004). Sensor networks closely interact with the users and their physical environments. Therefore, new security problems not happened before with other networks can emerge. Additionally, sensor network requires the protocol for group management so the group communication can be secure. The result of group computation is used for authentication to ensure that the group can be trusted. However, time and energy are the main concerns for any solution. In wired networks, the traffic and the computation resources are typically monitored and analyzed at some points. On the other hand, WSNs require a fully distributed and inexpensive solution while satisfying some requirements such as communication, energy, and memory. Particularly for the privacy, WSNs and wired networks have different threats. This difference causes the urgent needs not only technological responses but also the new laws.

It can be seen from above session that the security is critical for any security system including context-aware application. It is challenging to manage the security appropriately while there are many considerations such as privacy, confidentiality, data integrity, data availability, etc. needed to be taken account. As many context-aware applications involve a lot of wireless sensors and their networks, the concerns may differ from those of wired communication of devices. Although, there are some existing protection policies and mechanisms from all of the relevant communication portions, more detail of future study is required to discover the appropriate strategies for future context-aware applications. The future context-aware application will make the users less aware of their personal information which will be observed and collected unconsciously by the tiny wireless sensors around or on them. More detail of existing applications and some suggestion of future context-aware applications are explained in the next chapter.

References

Adams, A., & Sasse, M. A. (1999). Users are not the enemy. *Communications of the ACM, 42*(12), 40–46.

Albers, P., Camp, O., Percher, J. M., Jouga, B., Me, L., & Puttini, R. S. (2002, April). Security in Ad Hoc networks: A general intrusion detection architecture enhancing trust based approaches. In Wireless Information Systems (pp. 1–12).

Almutairi, S., Aldabbas, H., & Abu-Samaha, A. (2012). Review on the security related issues in context aware system. *International Journal of Wireless & Mobile Networks, 4*(3), 195.

Bardram, J. E., Kjær, R. E., & Pedersen, M. Ø. (2003, October). Context-aware user authentication —supporting proximity-based login in pervasive computing. In International Conference on Ubiquitous Computing (pp. 107–123). Berlin, Heidelberg: Springer.

Bell, D. (1996). The bell-lapadula model. *Journal of computer security, 4*(2), 3.

Burrows, M., Abadi, M., & Needham, R. M. (1989, December). A logic of authentication. In Proceedings of the Royal Society of London A: Mathematical, Physical and Engineering Sciences (Vol. 426, No. 1871, pp. 233–271). The Royal Society.

Carstens, D. S., McCauley-Bell, P. R., Malone, L. C., & DeMara, R. F. (2004). Evaluation of the human impact of password authentication practices on information security.

Devika, K., & Jawahar, M. (2015). Review on: Cryptographic algorithms for data integrity proofs in cloud storage.

Duncan, R. (2001). An overview of different authentication methods and protocols. SANS Institute.

Eissa, M. M., Ali, I. A., & Abdel-Latif, K. M. (2013). Wi-Fi protected access for secure power network protection scheme. *International Journal of Electrical Power & Energy Systems, 46*, 414–424.

Frank, M., Biedert, R., Ma, E., Martinovic, I., & Song, D. (2013). Touchalytics: On the applicability of touchscreen input as a behavioral biometric for continuous authentication. *IEEE Transactions on Information Forensics and Security, 8*(1), 136–148.

Giot, R., El-Abed, M., & Rosenberger, C. (2009, May). Keystroke dynamics authentication for collaborative systems. In Collaborative Technologies and Systems, 2009. CTS'09. International Symposium on (pp. 172–179). IEEE.

Guven, A., & Sogukpinar, I. (2003). Understanding users' keystroke patterns for computer access security. *Computers & Security, 22*(8), 695–706.

Harris, S. (2010). Access control. CISSP all-in-one exam guide (5th ed.), pp. 153–279.

Hickman, K., & Elgamal, T. (1995). The SSL protocol. *Netscape Communications Corp, 501.*

Ives, B., Walsh, K. R., & Schneider, H. (2004). The domino effect of password reuse. *Communications of the ACM, 47*(4), 75–78.

Juang, W. S. (2004). Efficient multi-server password authenticated key agreement using smart cards. *IEEE Transactions on Consumer Electronics, 50*(1), 251–255.

Kalita, H. K., & Kar, A. (2009). Wireless sensor network security analysis. *International Journal of Next-Generation Networks (IJNGN), 1*(1), 1–10.

Kim, H. J. (1995). Biometrics, is it a viable proposition for identity authentication and access control? *Computers & Security, 14*(3), 205–214.

Kumar, M. (2004). on the weaknesses and improvements of an efficient password based remote user authentication scheme using smart cards. *IACR Cryptology ePrint Archive, 2004,* 163.

Menkus, B. (1988). Understanding the use of passwords. *Computers & Security, 7*(2), 132–136.

Perrig, A., Stankovic, J., & Wagner, D. (2004). Security in wireless sensor networks. *Communications of the ACM, 47*(6), 53–57.

Rivest, R. L., Shamir, A., & Adleman, L. (1978). A method for obtaining digital signatures and public-key cryptosystems. *Communications of the ACM, 21*(2), 120–126.

Saevanee, H., & Bhatarakosol, P. (2008, December). User authentication using combination of behavioral biometrics over the touchpad acting like touch screen of mobile device. In Computer and Electrical Engineering, 2008. ICCEE 2008. International Conference on (pp. 82–86). IEEE.

Sasse, M. A., Brostoff, S., & Weirich, D. (2001). Transforming the 'weakest link'—a human/computer interaction approach to usable and effective security. *BT technology journal, 19*(3), 122–131.

Smith, R. E. (2002). The strong password dilemma. *Computer Security Journal, 18*(2), 31–38.

Snelick, R., Uludag, U., Mink, A., Indovina, M., & Jain, A. (2005). Large-scale evaluation of multimodal biometric authentication using state-of-the-art systems. *IEEE Transactions on Pattern Analysis and Machine Intelligence, 27*(3), 450–455.

Stallings, W. (2006). Cryptography and network security: Principles and practices. Pearson Education India.

Stallings, W., & Brown, L. (2008). Computer security. Principles and Practice.

Tuyls, P., Akkermans, A. H., Kevenaar, T. A., Schrijen, G. J., Bazen, A. M., & Veldhuis, R. N. (2005, July). Practical biometric authentication with template protection. In International Conference on Audio-and Video-Based Biometric Person Authentication (pp. 436–446). Berlin, Heidelberg: Springer.

Waichal, S., & Meshram, B. B. (2013). Router attacksdetection and defense mechanisms. *International Journal of Scientific & Technology Research, 2,* 145–149.

Wayman, J., Jain, A., Maltoni, D., & Maio, D. (2005). *An introduction to biometric authentication systems* (pp. 1–20). London: Springer.

Yan, J. J., Blackwell, A. F., Anderson, R. J., & Grant, A. (2004). Password memorability and security: Empirical results. *IEEE Security and Privacy, 2*(5), 25–31.

Chapter 6
Context-Aware Middleware and Applications

Abstract This chapter focuses on reporting some context-aware middleware and applications impacting the variety of application domains of context-aware computing for decades so that the readers will be able to utilize this kind of application appropriately. However, the main focus is on smart environments such as smart home, and personalized environments. At the end of this chapter, the future trends in smart environments for healthcare and education domains are suggested based on the evolution of existing works and some significant concerns.

To complete this book, it is necessary to demonstrate the existing middleware and applications of context-aware computing. This chapter aims to introduce the mian responsible application which is the smart environment. Especially, selecting the right middleware is crucial for non-technical developers. Besides understanding their applications precisely, the readers should also know how to choose the appropriate middleware. Consequently, some middleware are introduced in this chapter to provide the initial idea for selection. The future trend will be discussed based on the evolution of existing works. As a result, the reader will be able to predict the future trend of applications by taking some emergent types of context into account.

6.1 Context-Aware Middleware

As mentioned before, middleware obtains much attention from the developer for developing context-aware applications. It can accommodate the developer to concentrate on the application instead of putting much affords to physical communication with a large variety of sensors. Additionally, a variety of application areas can be customized easily by providing the general set of operating tools which are necessary for each single application. Middleware is the structure to collect context

P. Temdee and R. Prasad, *Context-Aware Communication and Computing: Applications for Smart Environment*, Springer Series in Wireless Technology, DOI 10.1007/978-3-319-59035-6_6

information, support the deployment of sensors and hide heterogeneity (Alegre et al. 2016). More specifically, it is a software layer between the network operating system and the applications. It promotes solutions to deal with many significant problems such as heterogeneity, interoperability, security, dependability, scalability, etc. From the literature (Perera et al. 2014), many existing middleware have shared some common components such as modeling, reasoning, dissemination method, etc. At the same time, all middleware also highlight some unique features to promote efficient context management frameworks according to their concerns and targeted applications. Although many middleware have been presented, a general purpose middleware applicable to any application still challenges the current and the future research. In this section, some existing middleware are introduced and discussed to conclude the trend of future research in context-aware middleware.

6.1.1 Existing Context-Aware Middleware

This section examines some existing context-aware middleware that can reveal the evolution of technology required for new applications. Each middleware is mainly designed to accomplish some particular objectives of individual applications. However, many of them typically share something in common. The selection of context-aware middleware still depends on applications.

6.1.1.1 Context Toolkit

Context Toolkit (Dey et al. 2001) is one of the early research works in providing the framework to support developers by containing some necessary features and abstractions. It can be considered as the solution having the architecture based on loosely coupled major components that interact among each other. Three main abstractions are introdiced including context widget for retrieving data from sensors, context interpreter for reasoning sensor data using different reasoning techniques, and context aggregator for creating higher level context. Context Toolkit identifies the standard features required by context-aware applications as capturing and accessing of context, context storage, context distribution, and context execution. Later, the context reasoning is supported by more decision models. It uses key-value modeling for context modeling. Context Toolkit is widely employed in many applications (Dey 2000). For example, an information display board shows the information to a user in front of it. An augmented whiteboard in the office shows messages for the staffs. A context-aware mailing list sends an incoming email only to staff who are currently in the office. A conference assistant aids the users when attending a conference. The CybreMinder (Dey and Abowd 2000) supports the creation, delivery, and handling of reminders. The smart home applications (Kidd et al. 1999) provide autonomous services. The augmented wheelchair uses the context to improve word prediction for the users.

6.1.1.2 Aura

Aura (Garlan et al. 2002) is on distributed architecture focusing on user's tasks from all devices of applications. Aura allows the users to maintain continuity for their works when moving between different environments. It can adapt the ongoing computation for a particular environment under the dynamic resources. Aura uses markup scheme and rule-based method for context modeling and reasoning respectively. It is on distributed architecture supporting peer to peer interactions. Aura consists of four main elements including the context observer for collecting and sending context, the task manager for managing the user tasks, the environment manager for managing context suppliers and related service, and the context suppliers for providing context information to other components. It can be seen that Aura is mainly used for promoting user mobility (Sousa and Garlan 2002). Many applications supporting campus collaboration are built on Aura (Fortino and Trunfio 2014) such as the application allowing the user to know the teammate on the campus, the application allowing the users to share their ideas with each other through distributed blackboard, etc.

6.1.1.3 CARISMA

CARISMA or Context-Aware Reflective Middleware System for Mobile Applications (Capra et al. 2003) focuses on the dynamic environment as the mobile environment. For the mobile environment, many devices will be networked. This environment enables the construction of distributed applications. These applications have to adapt their behavior to respond the changes of context appropriately such as network bandwidth's variation, power consumption, connectivity, etc. The main aim of CARISMA is the adaptation or the reflection. More specifically, this middleware is designed for mobile computing. The principle of reflection is used to enhance the construction of adaptive and context-aware mobile applications. CARISMA is distributed architecture, and the context is stored as application profile with XML-based language. It uses markup scheme and rule-based method for context modeling and reasoning respectively. For adaptation, it employs reflection to change middleware behavior which is dynamically modified by the application. For reflection model, the mobile devices can change operating context rapidly. The middleware is used for monitoring these changes. The context configuration determines the policies for a particular context. Finally, the reflective model allows the applications to change middleware behavior dynamically. Example applications built on CARISMA are conference applications such as talk reminding, messaging service, proceedings accessing, etc.

6.1.1.4 CoBrA

CoBrA or Context Broker Architecture (Chen et al. 2004) is an agent architecture called broker-centric architecture. It supports smart space application by providing knowledge sharing and context reasoning. CoBrA has component based architecture and uses Web Ontology Language (OWL) to define ontologies for context modeling and reasoning. A rule-based method is used to interpret and reason about context. A policy language and engine are used to control the sharing of user context. The primary element of CoBrA is a context broker. This broker will sense and reason about context based on the capability of the agents. The agents will share contextual knowledge, protect user privacy and maintain the consistent of contextual knowledge. More specifically, the context broker consists of four functional components including the context knowledge for providing persistent storage, the context reasoning engine for reasoning over context information, the context acquisition module for retrieving the context from many sources, and the policy management module for managing policies. The typical applications of CoBrA are the applications for intelligent meeting rooms.

6.1.1.5 SOCAM

SOCAM or Service Oriented Context-Aware Middleware (Gu et al. 2005) is ontology based middleware. SOCAM is mainly designed to support semantic representation, context reasoning, and knowledge sharing. It uses ontology-based method for both context modeling and context reasoning. A rule-based method is used for context interpretation. Its architecture is service-based architecture where many services are working together instead of having central control component. Two levels of ontologies are used including the lower-level and upper-level ontologies. While the lower level ontology deals with domain-specific descriptions, the upper-level ontology deals with general concepts. SOCAM composes of Context Providers, Context Interpreters, Context Database, Location Service, and Context-aware Mobile Services. Firstly, the Context Providers provide context information and represent it as context events with OWL descriptions. The Context Interpreters are responsible for high-level context information. The Context Interpreters include Context Reasoners containing rules for triggering the appropriate actions regarding the changes of context, and the Context Databases containing all instances of the ontology. The Location Service locates the context providers. Finally, the mobile services are applications and services that can adapt their behavior according to the context information. SOCAM is well-known for smart home environment.

6.1.1.6 e-SENSE

e-SENSE (Gluhak et al. 2006) aims to provide the technology for capturing the desired ambient intelligence surrounding the users and the service through a

Wireless Sensor Network (WSN) environment. The WSN consists of several sensor nodes which sense data of a physical phenomenon and communicate with each other over a radio link. The richness of contextual information that is required to capture ambient intelligence demands a multitude of multi-sensory information entirely. More specifically, e-SENSE enables ambient intelligence by using wireless multi-sensor networks for making context information available to the applications and the services. e-SENSE has distributed and node based architectures which promote peer to peer interactions. It allows the deployment of the software to communicate and process data in sensor networks. It uses rule-based method for context reasoning and has the functions relating to security and privacy. For IoT applications, three main components work together including body sensor networks (BSN), object sensor networks (OSN), and environment sensor networks (ESN). The e-SENSE protocol stack architecture is divided into four logical subsystems including the application (AP), management (MA), middleware (MI), and connectivity (CO) subsystem. The AP hosts one or several sensor applications. Every application can send and receive sensor data by using the services provided by MI. The MA is used for the configuration and initialization between MI and CO. The MA defines the role of all nodes in the sensor network. The MI provides some necessary services such as a data transfer service for delivering of the application data packets, the management service for executing node or service discovery, etc. The CO consists of all functions that are required for operating in some particular layers including the physical layer, the medium access control, the network, and the transport layer.

6.1.1.7 HCoM

HCoM or Hybrid Context Management (Ejigu et al. 2007) is a hybrid approach. It is the combination between semantic ontology and relational schema. The hybrid approach is required to deal particularly with a large size of data. HCoM deals with the standard functions including collection, organization, representation, storing, and presentation of the context. Therefore, it has the centralized architecture with five layers including acquisition layer, pre-processing layer, data modeling and storage layer, management modeling layer, and utilizing layer. Additionally, HCoM has some essential elements required for context management solution. For example, the context manager is required for aggregating the results and transmitting the obtained data to reasoning component. The collaboration manager is required for gathering more data from other context sources. The context filter is used for validating and making decision whether the context needs to be stored. The context selector is used for making decision what context should be used in reasoning processing. The database is also needed for keeping all rules for context manipulations. HCoM uses graphical and ontology-based method for context modeling. For context reasoning, the rule-based and ontology-based methods are used.

6.1.1.8 SIM

SIM or Sensor Information Management (Baek et al. 2007) employs an agent-based architecture to provide the framework for context-aware applications. It has the distributed architecture and aims to address location tracking in the smart home applications. It focuses on collecting sensor data from multiple sources and aggregating them to derive higher level information. The information of node and attribute levels is required. The contexts in node level are node ID, location, and priority. Attributes are context attribute and its corresponding measurement. A location tracking algorithm using a mobile positioning device has been introduced which is responsible by the position manager. SIM uses sensor priority to resolve the conflicts of sensor information. The context manager and decision component together with aggregation method are used for solving the conflict. SIM also uses key-value and graphical based methods for context modeling and rule-based method for context reasoning respectively.

6.1.1.9 COSMOS

COSMOS or COntext entitieS coMpositiOn and Sharing (Conan et al. 2007) is middleware mainly designed for ubiquitous environments. COSMOS has distributed and node based architecture promoting the scalability. It identifies the contextual situations and models it with the context policies that are hierarchically decomposed into the context nodes. The sharing and encapsulation are the most important relationships between context nodes. The context node consists of its activity manager, context processor, context reasoner, context configurator, and message managers. COSMOS employs user preferences as the context information. Therefore, COSMOS consists of three layers including context collector, context processing, and context adaptation. Context collector gathers information from the sensors. Context processing derives high-level information from raw sensor data. Context adaptation provides access to the processed context for the applications. The object-oriented method is used for context modeling, and rule-based method is used for context reasoning respectively. Additionally, COSMOS provides the predefined context operators to the developers such as elementary operators for collecting raw data, add operator for data merging, thresholds operator for data merging and abstraction, etc. Therefore, COSMOS can be considered as the user-friendly middleware.

6.1.1.10 Hydra

Hydra (Eisenhauer et al. 2010) refers to Networked Embedded System Middleware for heterogeneous physical devices in a distributed architecture. It is an IoT middleware for integrating wireless devices and sensors into ambient intelligence environments. It is one of the early efforts of IoT middleware focusing on connecting embedded devices to the applications. Hydra allows seamless access to the features of many devices without taking some important mechanisms into account such as its manufacturer, interfaces, location, and communication. It is a middleware based on a service-oriented architecture. It is mainly designed to support the distributed and centralized architectures, the reflective properties of the middleware, the security and the trust enabling components. The new applications are initiated with Hydra especially in facility management as smart homes, smart hospital, and smart farm, etc. It is expected to be the middleware that is adapted to new requirments while enhancing scalability, robustness, and security. Hydra demonstrates the importance of pluggable rules allowing insertions when necessary. Hydra has a Context-Aware Framework (CAF) to provide the capabilities of both high-level and lower-level semantic processing. CAF consists of two main components including Data Acquisition Component Context Manager. While the Data Acquision Component is responsible for connecting and retrieving data from sensors, the Context Manager is responsible for context management, context awareness, and context interpretation. CAF models three distinct types of context including the device contexts, the semantic context, and the application context. The example of device context is the data source. The examples of semantic contexts are location, environment, etc. Hydra uses several context modeling methods including key-value, ontology based and object based methods. Rule-based and ontology-based methods are used for modeling. There are some functions relating to security and privacy appeared in Hydra.

6.1.1.11 Feel@Home

Feel@Home (Guo et al. 2010) is a context management framework for supporting the interaction between different domains. Since context information is stored with OWL, the ontology-based method is used for context modeling and reasoning. The graphical method is also used for context modeling. Feel@Home supports two different interactions including the intra-domain and the cross domain which are essential for the IoT paradigm. Sensor networks usually deal with one domain while IoT is required to deal with multiple domains. Feel@Home aims to fulfill this requirement as context management framework for IoT paradigm with distributed and node based architecture. It consists of three parts including user queries, global administration server (GAS), and domain context manager (DCM). User queries decide what the relevant domain is involved in answering the user query. GAS redirects the user query to the relevant context managers. DCM consists of standard context management components such as context wrapper for gathering context from sensors and other sources, context aggregator for triggering context reasoning, context reasoning, etc. Feel@Home has been demonstrated in three different application domains including smart home, smart office, and mobile application with some functions relating to security and privacy.

6.1.1.12 Octopus

Octopus (Firner et al. 2011) is an open-source and extensible system that supports data management and IoT applications. There have been many IoT applications nowadays. Unfortunately, it is difficult for the non-technical people to deploy and use these applications. The widespread adoption of these applications is thus distracted by the high-level of technical requirements. Therefore, Octopus is designed to help non-technical people to deploy sensors, manage context, and develop their applications quickly. More specially, Octopus develops middleware abstractions and programming models for non-technical developer to easily develop IoT applications. Octopus has distributed and node based architecture. Octopus abstraction layers separate the developer from performing data analysis and manipulating in the application. The Aggregator distinguishes the sensing layer from data analysis and allows the application to support multiple sensing layers seamlessly. Octopus is widely used on the smart home and office domains.

6.1.1.13 SCIMS

SCIMS or Social Context Information Management System (Kabir et al. 2012) is a social network middleware having centralized architecture. SCIMS allows efficient access to social context information while respecting the privacy of the user. It also aims to facilitate the development of social-aware applications. It defines social context as a set of interaction information among the users. Three major contributions are found in SCIMS. Firstly, an ontology-based method is used for representing and keeping both relationships and user's status information. The relationship information covers people and object-centric social relationship. Consequently, the ontology-based method is used for social context modeling and reasoning. Secondly, social context information can be derived from multiple sources. Finally, the owner privacy can be preserved by fine-tuning the granularity of information access accordingly to the control policies. SCIMS uses semantic web technologies to implement the overall system. It comprises two layers including information acquisition and information management layers. The information acquisition layer is used for collecting social context from various sources such as calendar entries, physical sensors, etc. At the same time, SCIMS has an interface to acquire social data from all sources. For information management layer, an ontology-based context model is used to store the social data, and provide more complex context information. For controlling access to SCIMS, the policy model reflecting the thinking way of human is used. Finally, the query interface is employed so that the application developers can quickly develop applications without dealing with complicated data representation and management.

6.1.1.14 CAMEO

CAMEO or Context-Aware Advertising Mediator and Optimizer (Khan et al. 2013) is a framework for mobile advertising. It is developed for making the advertisement applications to be more consumer-friendly and intelligent. CAMEO is considered as the social context middleware. The social context is defined as the information derived from both virtual and physical social interactions among the users. It uses object-role model for social context modeling and rule-based method for social context reasoning. Besides social tie inference, CAMEO also focuses on group detection on distributed architecture. The group is defined as some users who are similar to each other regarding some particular attributes such as interest, habits, etc. Moreover, the group dynamics is detected by analyzing the evolution of social interactions over time. CAMEO also uses context prediction to reduce the bandwidth and energy consumption. CAMEO has three steps that mediate the interactions among individual applications. Firstly, it takes a corpus of advertisements from multiple Advertisement Networks (ANs). Secondly, it serves advertisements from the stored corpus to the mobile application. Thirdly, it negotiates with the Internet Service Providers (ISPs). Four principal components are used in CAMEO including the Context Predictor, the Advertisement Manager, the ISP Negotiator, and the Accounting and Verification module. The advertisement pre-fetching and local serving functionalities are provided by the Context Predictor and the Advertisement Manager components. The ISP Negotiator performs the functionality of bartering advertisement privileges for accessing bandwidth. Each of these three components interacts with the Accounting and Verification module for assuring that the local advertisements are being served correctly.

6.1.2 Concerns of Context-Aware Middleware

As shown above, some examples of existing middleware have shared some common components. For context modeling and reasoning methods, the same approach is frequently used. Some middleware use the same type of context modeling and reasoning method such as Context Toolkit, Aura, CARIAMA, COSMOS, etc. while some middleware require combinations of several methods as SOCAM, HCoM, Hydra, etc. For architectures, some middleware have unique architecture as Aura, CARISMA, SOCAM, HCoM, SIM, Hydra, etc., while some have the hybrid architecture such as Context Toolkit, e-SENSE, COSMOS, Feel@Home, Octopus, etc. The development trend of context-aware applications also depends on the emergent technology. Some middleware are developed to serve the requirements explicitly from the user demands together with the supporting technology. For example, CARISMA and e-SENSE are mainly designed for mobile computing environment while Hydra, Feel@Home, and Octopus are specially designed for IoT environment. It can be seen clearly that almost middleware are required to deal with a variety of sensors in sensor networks. The trend implies current middleware should be designed in the way that it will be easy deployed by non-technical people. Many of existing middleware aim to promote the facility management such as the smart home, the smart office, and the mobile application. Many of current middleware seek to encourage the utilization of social context in many application domains such as business domain (Khan et al. 2013), public security (Yu et al. 2012), public health (Eubank et al. 2004), etc. Social-aware applications need to gather and process various data over heterogeneous software and hardware components to obtain proper social context. This task brings critical challenges to the application developers. The social-aware middleware should also offer inference services of different social contextsto fully support social context-aware applications. More specifically, it should provide services of social context inference in various levels (Liang and Cao 2015). Therefore, some significant concerns regarding the future trend of middleware should be the availability of generic framework to deal with a variety of sensors, the flexible manipulation of variety kinds of contexts especially the social context, and the ease development tool for the non-technical people. Therefore the adoption of context-aware applications can be widely implemented. Additionally, the issues of security should also be the main concern for the future development of context-aware middleware, especially in IoT application.

6.2 Context-Aware Applications for Smart Environment

For this book, the context-aware application is considered to be the application that can act between some degrees of rationally and personally with or without the intervention from the users. The applications mentioned in this chapter aim to promote the applications of context-aware computing particular in the smart environment for both physical and virtual environment. In Chap. 1, some new application areas have been introduced. For this chapter, the applications particularly for smart environments will be discussed.

6.2.1 Smart Home

The smart home application involves various kinds of supporting technologies not only context-aware computing but also other emergent technologies such as ubiquitous computing, wireless sensor technology, wireless communication, embedded technology, ambient intelligence, etc. At the early stage, the first definition of smart homes was provided by Lutolf (1992) that "the smart home concept is the integration of different services within a home by using a common communication system". Many definitions were later proposed to include more idea of smartness or intelligence not only automation (Allen et al. 2001; Briere 2011). Following this definition, much attention pays to more involvement of the users (Satpathy 2006). Nowadays, a smart home can be seen as one of the most popular applications in which the home environment is monitored by ambient intelligence to provide context-aware services and facilitate remote home control (Alam et al. 2012).

The main objective of smart home applications is to embed smartness into the house for some specific purposes such as comfort, safety, security, energy conservation, etc. Due to the advance of telecommunication and web technology, remote and local monitoring systems (Schlager and Baringer 1997; Petite and Huff 2004) become common and affordable components of smart home applications. Additionally, the local and remote control systems can be found in healthcare purpose (Mihailidis et al. 2004; Virone et al. 2002; Farella et al. 2010) such as patient monitoring and senior people monitoring. For this purpose, variety kinds of contexts are used such as motion, temperature, image, voice, video, etc. to cooperate with associated applications. Multi-sensors systems are employed for gathering contexts by accessing directly to the sensors or through the platforms which are designed to cooperate all sensors together. Smart hospital at home is another application extending from the smart home application and gaining much interest nowadays. Currently, many countries all over the world are entering into aging society. Their people are getting old with high possibility to have health problem, especially chronic diseases. Moreover, there are also the increasing numbers of disability people and patients who requires long term treatments. For these reasons, smart hospital at home seems to be the only appropriate solution that can cooperate with assistive technologies to accommodate healthcare services at home.

Together with IoT technology, the smart home applications can enhance the quality of life by introducing automated appliance control and assistive services (Darianian and Michael 2008; Gubbi et al. 2013; Soliman et al. 2013). IoT illustrates the employment of context-aware computing explicitly by storing the context information which is linked to sensor data, interpreting them and making them be more meaningful and understandable to perform machine to machine communication. Although IoT can make home appliances enable to communicate with each other including the users and offer them a better quality of life, some important issues especially in the privacy and security issues require more intensive consideration.

6.2.2 Personalized Environments

This section mainly focuses on two application domains including healthcare and education. The evolution of personalized environments for both areas is also discuss in this section.

6.2.2.1 Smart Healthcare Environments

A significant shift in healthcare has taken because of the advance in Information Communication Technology (ICT). For example, ICT can help decreasing demands of paper usage as it can be found that the medical work is accomplished by a wide range of documents, schemas, charts, etc. It can provide the useful communication tools for solving the interruptions of direct or indirect communication ranging from laboratory results to complex consultation and advice (Coiera 2000; Spencer and Logan 2002).

Current technologies have allowed the introduction of awareness system for healthcare activities such as wireless technologies, mobile technology, sensors technology, wearable instruments, handheld computers, etc. Such technologies could help enhancing the quality of care because they can help professionals to manage their tasks appropriately and efficiently. In particular, some research works have underlined that the context aware applications promote the successful tools for cooperative work among healthcare professionals (Bricon-Souf et al. 1999; Renard et al. 1999; Reddy et al. 2002). It can be seen that ICT has provided the environment for exchanging ideas, information and knowledge to solve cooperative problems benefiting for both caregivers and their patients. The cooperation between health care professionals can be more efficiently by being mediated through variety kinds of digital platforms and mobile tools (Bricon-Souf et al. 2003). Electronic Health Record (EHR) is the electronic form of patient's health data. Now, there is the requirement for healthcare professionals to access EHR not only on the desktop computer but also on any smart portable devices as smartphones or tablets (Gans et al. 2005; Grasso 2004; Bricon-Souf and Newman 2007). Many context-aware applications have shown the favorable personalized views on the patient's EHR at some point relevant to the current situation and mobility support. With the advance of wireless communication and sensor technology, context-aware computing also has introduced the system for patients monitoring for both at home and the hospital. The physical sensors together with patient's personal records are used as the crucial contexts for monitoring (Kim et al. 2014; Gelogo et al. 2015). Together with the mobile application, the context-aware applications typically can provide better diagnosis and treatment services not only monitoring service. Because of the shortage numbers of caregivers especially in the countries facing aging society, the context-aware mobile applications are very useful not only for the chronic patients but also the senior who are regularly stay at home without the intensive care from the caregivers. For the modern healthcare, the context-aware applications are considered as the proper tools to promote personalized information not only for the patients, chronic patients, or senior people but also for any people who concern about their health. The physical sensors together with their lifestyle information are used as the necessary context triggering the personal healthcare assistants or recommendation systems accordingly to the user's preferences and their health conditions (Chawla and Davis 2013; Simmons et al. 2012; Golubnitschaja and Costigliola 2010). As it can be seen that context-aware applications have been demonstrated significant impact to healthcare application domain for decades, this area is still appealing for many researchers to work on because it will help enhancing the quality of life of people that will significantly impact the socio-economics development for all countries. However, there are also some significant concerns about developing this kind of applications in healthcare domains which are worth to take intensive consideration especially the security and privacy issues.

6.2.2.2 Smart Learning Environments

For decades, ICT also has been playing the main role for the paradigm shift in the education area. Many existing research works have been demonstrating the evolution of this area. The trend of the learning supporting system (Cheng et al. 2005) shows that the innovations for education domain rely on the support technologies emerging at different points of time. Computer Aid Instruction (CAI) and Instruction Tutoring System (ITS) can be considered as the shifting pioneers of the education system. The learners can learn by themselves with the stand-alone programs (Arnold 1997; Graesser et al. 2005) without face to face learning with their teacher as in the traditional classroom. They are widely used since the first emergence of the personal computer. Later, electronic learning (e-learning) has been widely introduced due to the advance of computer communication and the introduction of Internet technology. At the early stage, e–learning aims to transform the learning content into the digital form, deliver it and make it available to a large number of learners anywhere and anytime (Rosenberg 2001; Zhang et al. 2004). Later, e-learning aims to provide the virtual learning environment having space and communication tools the same as those in the traditional classroom to promote effective communications between the teachers and the learners and all among the learners as illustrated by many works in computer supported collaborative learning area (O'Malley 2012; Lipponen 2002). At the same time, the Web Based Learning (WBL) is widely introduced to support e-learning (Chumley-Jones et al. 2002; Zaıane and Luo 2001). With the advance in web technology, WBL has gained interests dramatically from many researchers not only to promote anywhere anytime learning but also interactive learning (Johnson et al. 2000). After mobile learning (m-learning) is widely announced since the mobile technology and hands-on devices are affordable, the learning system has been shifted again to be more accessible and personalized learning (Winters 2007). Nowadays, it is the era of ubiquitous learning (u-learning) since ubiquitous computing together with the wireless sensor network, and embeded technology has been widely introduced and implemented (Barbosa et al. 2008; Hwang and Tsai 2011; Hwang 2006; Temdee 2014). U-learning has been presented respectively because of the advent of ubiquitous computing (Weiser 1991), where the computers are blended into our everyday lives inconspicuously. With many advanced technologies, u-learning is considered as the setting of new education system in which the learning process happens all around the learners unconsciously. With IoT technology, the u-learning becomes even more conscious with the obtrusivly support from multi-sensors surrounded by or on the learners. This learning enviornment promotes the freedom to learn for the learners.

Along with the evolution of technology supporting learning mentioned before, the concept of context-aware computing has already integrated implicitly and explicitly into those learning systems. Context-aware applications can be found with an enormous diversity in education domain covering from the stand-alone application as automatic tutoring system to the web-based, mobile and ubiquitous learning environments. Generally, context awareness is required by many educational systems. For example, the assessment of the learner's ability (Boud and Brew 1995; Stone 2014; Ihantola et al. 2010) as commonly seen in the recommendation systems (Verbert et al. 2012) and personalized learning systems (Klašnja-Milićević et al. 2011; Dolog et al. 2004; Munoz et al. 2004) that can provide the appropriate feedbacks, recommendations or interventions to the individual learner. In the recent years, several works on context-aware education (Das et al. 2010; Beale and Lonsdale 2004; Berri et al. 2006) have been proposed to accomplish the appropriate personal response for the learners with different kinds of context such as personal profile, learning style, preference, etc. However, those works still leave some space for the researchers to discover new findings to support active learning.

6.3 Future Context-Aware Applications

This section suggests the future trend of context-aware applications based on the different points of views on social context that can make the application being aware more rationally and personally. As mentioned before, the definition of social context should not only focus on the interactions among people but also among machines and even between the people and the machine. The machine can be any computing entity enabling to communicate with each other or with other entities such as sensors, software agents, applications, etc. It is easy to have these computing entities nowadays because of the IoT technology. As mentioned before, the social networks aiming to understand and measure the interaction among people has been contributing in many useful social-aware applications nowadays. It is also applicable to explain the interactions and create the social contexts of other computing entities. Also, the fusion of those contexts may also provide great benefit to social context definition. Therefore, social-aware applications in the future are suggested to have more extension of social context definition as the conceptual diagram is shown in Fig. 6.1. It can be seen from Fig. 6.1 that while the people are interacting with each other, other interactions happen at the same time such as the interactions between people and the machine, the interactions among the machines, etc. The machines can be physical or virtual entities.

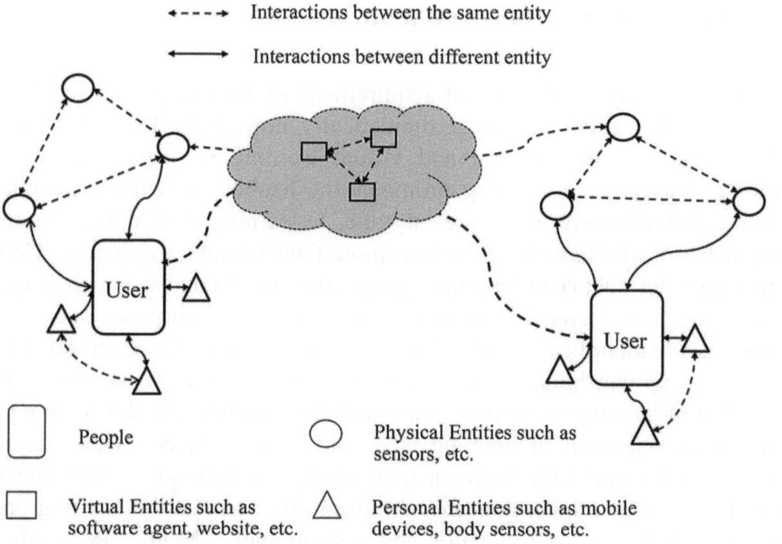

Fig. 6.1 Conceptual diagram for social-aware application

6.3.1 Future Social-Aware Applications

Research in social-aware applications has also provided the significant benefits to the area of business, especially in the digital marketing. Social networks and patterns of their relationships are necessary information for the analysis of marketing behavior which will be used effectively for marketing strategies. A major challenge facing marketing strategists is how to increase the market size by using social network based marketing strategies. To accomplish this goal, the collected social network related data as a social context has to be analyzed appropriately. The most difficult part comes from the data acquisition process which is usually done manually in the early stages. Although the online data acquisitions are available later on, the appropriate data classification methods are indeed required for some particular applications. This task requires the intelligent algorithm for dealing with the big data. Currently, the digital marketing becomes bigger and bigger because of the convenient access to the online market. The people sometimes do not realize that they are communicating with the software agent acting in different roles instead of the human such as the brokers, the dealers, and the distributor, etc. The communication among these software agents may help better understanding of the market behavior in the different points of view and may impact for the social-aware application in the future.

6.3.2 Future Education Applications

For education domain, the urgent requirement of having personalized learning environments has been increasing dramatically nowadays. These learning environments can be both physical and virtual learning workspace as also called u-learning. This kind of learning promotes the learners to achieve their learning outcomes unconsciously anywhere anytime. U-learning is designed for the 21st century learners who have freedom to explore their learning experience. Generally, for any computer enhanced learning system, the social context which is the interactions among the learners is typically used for many purposes such as group monitoring, role identification, etc. This social context is still required for u-learning where the group can be small as in the class or bigger as in the social network society. For u-learning, the most important components are the learning object which can be appeared in both physical and virtual objects. These objects are designed to understand the learners individually by their preference and performance. They can also communicate with each other so that the learning paths are given to each learner individually and appropriately. More specifically, these behaviors can be considered as the social context that can be later used to provide appropriate response or intervention to the learners based on their individual contexts. The learners are evaluated and modeled by using their contexts so that the learners can achieve their learning goals with their learning paths. Another kind of social context is also crucial especially the interaction among these learning objects. They can be used for verifying the contents of each learning objects appropriately. Consequently, they can help planning for the appropriate contents for each learning object. Some contents can be separately kept in other learning objects to satisfy load balancing and time consumption. It can be seen that the extension of social context, which is not only the interactions among the learners, can potentially provide more benefits for personalized learning which is a popular application of social contexts in the education domain.

6.3.3 Future Healthcare Applications

For healthcare domain, it is obvious to see from the current smart healthcare services that all smart physical objects and virtual objects are working together to provide the efficient healthcare services for the patients and the ordinary people. The smart hospital and smart healthcare mobile services are famous examples. Nowadays, there are many senior and disability people who have to stay alone at home. As mentioned before, the requirement to have the smart hospital at home is increasing dramatically. The smart hospital at home just simply means the home that the individual can receive the same care services as those they can have in the hospital. However, they can feel more convenient and comfortable at home. This system requires not only the personal context and the environment context but also the body context and the social context to fulfill the sucess of this application. The body context is gathered from multiple body sensors which can be facilitated with wireless communication. The social context among these body sensors and the sink or even the manager nodes can be used for predicting the health condition of the individual. Additionally, the social context which is the interaction among people becomes critical now. This social context can help the individual get support from their community. It is another way of treatment that affects their quality of life. In conclusion, the healthcare application domain is still considered as the important contribution area of context-aware application.

6.3.4 Suggestion for Future Context-Aware Applications

In term of information management, the social context which is the interaction among people provides many benefits for many application areas such as education, public health, business, etc. These applications require the combination of different types of contexts to obtain more useful application. For example, the health department can monitor the spread of infectious disease by location-based group information. The security department can perform crowd detection and criminal analysis by group location. The urgent requirement of these social-aware applications is to provide real-time system adaptation because the social contexts are dynamic. Additionally, the social context which is the interactions among people and virtual objects such as websites, services, etc. can help providing more useful context for analyzing.

Context-aware applications still have high potential to provide the contribution to many application domains. The future context-aware applications will focus on both the applications act personally and rationally to satisfy the users and the environments. The combination of context types is required for future context-aware applications especially the social context which represents the interaction information between any entities not only among people. The communications among entities will be seamless and less conspicuous. Moreover, the need for security issues becomes more crucial. It is strongly required for any application. More importantly, the users themselves need to be engaged to be part of the context-aware application. It is not easy to engage the users to trust these smart applications. However, it is worth to convince the users to utilize this kind of application to satisfy their daily life at least to some degree. This requirement is still challenging and needs more research works to fulfill.

As mentioned before, the utilization of context-aware computing is gaining much interest nowadays. The new types of context have been introducing novel applications in many application areas. The most concerns of using context-aware applications are not only how to acquire the complete contexts from different kinds of sensors, but also how to make senses from all those detected information. Moreover, the communications are necessary for context-aware applications. The communication not only is essential for perceiving component but also for thinking and acting components. All communication types need to be carefully designed so that there will not be any data loss and damage on the way of transmission. Like any other applications, the security issues become crucial for context-aware application nowadays. For context-aware applications, the security issues become even more challenging because the users will have to deal with the machines that can interact with them obtrusively all the time. They do not even realize how their personal or confidential data will be utilized. However, the users always require the appropriate responses from the application while they don't want their personal information to be revealed. This evidence still opens enormous opportunity for the future study to discover the method how to satisfy the individual need of each single user of context-aware applications.

In conclusion, this book provides all necessary elements for the readers who would like to apply context-aware applications to their application domains. The basic knowledge contained in this book is expected to guide the readers to be familiar with context-aware computing. Although the context-aware applications can be the optimal solutions for the personalized and rationalized requirements, designing and developing this kind of application require additional considerations as mentioned throughout this book. Additionally, the future study and research in context-aware computing is still challenging to discover new findings and extend its contribution to wider application domains.

References

Alam, M. R., Reaz, M. B. I., & Ali, M. A. M. (2012). A review of smart homes—past, present, and future. IEEE Transactions on Systems, Man, and Cybernetics, Part C (Applications and Reviews), *42*(6), 1190–1203.

Alegre, U., Augusto, J. C., & Clark, T. (2016). Engineering context-aware systems and applications: A survey. *Journal of Systems and Software, 117,* 55–83.

Allen, B., van Berlo, A., Ekberg, J., Fellbaum, K., Hampicke, M., & Willems, C. (2001). Design guidelines on smart homes (No. LEMA-BOOK-2001–004). COST 219bis guidebook.

Arnold, D. N. (1997). Computer-aided instruction. Encarta Encyclopedia. Microsoft.

Baek, S. H., Choi, E. C., Huh, J. D., & Park, K. R. (2007). Sensor information management mechanism for context-aware service in ubiquitous home. *IEEE Transactions on Consumer Electronics, 53*(4), 1393–1400.

Barbosa, J., Hahn, R., Rabello, S., & Barbosa, D. (2008). Local: A model geared towards ubiquitous learning. *ACM SIGCSE Bulletin, 40*(1), 432–436.

Beale, R., & Lonsdale, P. (2004, September). Mobile context aware systems: The intelligence to support tasks and effectively utilise resources. In *International Conference on Mobile Human-Computer Interaction*, (pp. 240–251). Berlin Heidelberg: Springer.

Berri, J., Benlamri, R., & Atif, Y. (2006, July). Ontology-based framework for context-aware mobile learning. In *Proceedings of the 2006 International Conference on Wireless Communications and Mobile Computing*, (pp. 1307–1310). ACM.

Boud, D., & Brew, A. (1995). Developing a typology for learner self assessment practices. *Research and development in Higher Education, 18,* 130–135.

Bricon-Souf, N., Renard, J. M., & Beuscart, R. (1999). Dynamic workflow model for complex activity in intensive care unit. *International Journal of Medical Informatics, 53*(2), 143–150.

Bricon-Souf, N., Dufresne, E., Beuscart-Zephir, M. C., & Beuscart, R. (2003). Communication of information in the homecare context. The New Navigators: From Professionals to Patients: Proceedings of MIE2003, 95.

Bricon-Souf, N., & Newman, C. R. (2007). Context awareness in health care: A review. *International Journal of Medical Informatics, 76*(1), 2–12.

Briere, D. (2011). Smart homes for dummies. Wiley.

Capra, L., Emmerich, W., & Mascolo, C. (2003). Carisma: Context-aware reflective middleware system for mobile applications. *IEEE Transactions on Software Engineering, 29*(10), 929–945.

Chawla, N. V., & Davis, D. A. (2013). Bringing big data to personalized healthcare: a patient-centered framework. *Journal of General Internal Medicine, 28*(3), 660–665.

Chen, H., Finin, T., Joshi, A., Kagal, L., Perich, F., & Chakraborty, D. (2004). Intelligent agents meet the semantic web in smart spaces. *IEEE Internet Computing, 8*(6), 69–79.

Cheng, Z., Sun, S., Kansen, M., Huang, T., & He, A. (2005, March). A personalized ubiquitous education support environment by comparing learning instructional requirement with learner's behavior. In *19th International Conference on Advanced Information Networking and Applications (AINA'05) Volume 1 (AINA papers)* (Vol. 2, pp. 567–573). IEEE.

Chumley-Jones, H. S., Dobbie, A., & Alford, C. L. (2002). Web-based learning: Sound educational method or hype? A review of the evaluation literature. *Academic Medicine, 77*(10), S86–S93.

Coiera, E. (2000). When conversation is better than computation. *Journal of the American Medical Informatics Association, 7*(3), 277–286.

Conan, D., Rouvoy, R., & Seinturier, L. (2007, June). Scalable processing of context information with COSMOS. In *IFIP International Conference on Distributed Applications and Interoperable Systems*, (pp. 210–224). Berlin Heidelberg: Springer.

Darianian, M., & Michael, M. P. (2008, December). Smart home mobile RFID-based Internet-of-Things systems and services. In *2008 International Conference on Advanced Computer Theory and Engineering*, (pp. 116–120). IEEE.

Das, M., Bhaskar, M., Chithralekha, T., & Sivasathya, S. (2010). Context aware e-learning system with dynamically composable learning objects. *International Journal on Computer Science and Engineering, 2*(4), 1245–1253.

Dey, A. K. (2000). Providing architectural support for building context-aware applications (Doctoral dissertation, Georgia Institute of Technology).

Dey, A. K., & Abowd, G. D. (2000, September). CybreMinder: A context-aware system for supporting reminders. In *International Symposium on Handheld and Ubiquitous Computing*, (pp. 172–186). Berlin Heidelberg: Springer.

Dey, A. K., Abowd, G. D., & Salber, D. (2001). A conceptual framework and a toolkit for supporting the rapid prototyping of context-aware applications. *Human-computer Interaction, 16*(2), 97–166.

Dolog, P., Henze, N., Nejdl, W., & Sintek, M. (2004, August). The personal reader: Personalizing and enriching learning resources using semantic web technologies. In *International Conference on Adaptive Hypermedia and Adaptive Web-Based Systems*, (pp. 85–94). Berlin Heidelberg: Springer.

Eisenhauer, M., Rosengren, P., & Antolin, P. (2010). Hydra: A development platform for integrating wireless devices and sensors into ambient intelligence systems. In *The Internet of Things*, (pp. 367–373). New York: Springer.

Ejigu, D., Scuturici, M., & Brunie, L. (2007, October). Semantic approach to context management and reasoning in ubiquitous context-aware systems. In *Digital Information Management, 2007. ICDIM'07. 2nd International Conference*, (Vol. 1, pp. 500–505). IEEE.

Eubank, S., Guclu, H., Kumar, V. A., Marathe, M. V., Srinivasan, A., Toroczkai, Z., et al. (2004). Modelling disease outbreaks in realistic urban social networks. *Nature, 429*(6988), 180–184.

Farella, E., Falavigna, M., & Riccò, B. (2010). Aware and smart environments: The Casattenta project. *Microelectronics Journal, 41*(11), 697–702.

Firner, B., Moore, R. S., Howard, R., Martin, R. P., & Zhang, Y. (2011, November). Poster: Smart buildings, sensor networks, and the internet of things. In *Proceedings of the 9th ACM Conference on Embedded Networked Sensor Systems*, (pp. 337–338). ACM.

Fortino, G., & Trunfio, P. (2014). *Internet of things based on smart objects*. Cham: Springer International Publishing.

Gans, D., Kralewski, J., Hammons, T., & Dowd, B. (2005). Medical groups' adoption of electronic health records and information systems. *Health Affairs, 24*(5), 1323–1333.

Garlan, D., Siewiorek, D. P., Smailagic, A., & Steenkiste, P. (2002). Project aura: Toward distraction-free pervasive computing. *IEEE Pervasive Computing, 1*(2), 22–31.

Gelogo, Y. E., Kim, H. K., & Jung, R. (2015). Context-aware computing for delivering u-Healthcare services. *International Journal of Smart Home, 9*(8), 169–178.

Golubnitschaja, O., & Costigliola, V. (2010). Common origin but individual outcomes: Time for new guidelines in personalized healthcare. *Personalized Medicine, 7*(5), 561–568.

Graesser, A. C., Chipman, P., Haynes, B. C., & Olney, A. (2005). AutoTutor: An intelligent tutoring system with mixed-initiative dialogue. *IEEE Transactions on Education, 48*(4), 612–618.

Grasso, M. A. (2004, June). Clinical applications of handheld computers. In *Computer-Based Medical Systems, 2004. CBMS 2004. Proceedings. 17th IEEE Symposium*, (pp. 141–146). IEEE.

Gubbi, J., Buyya, R., Marusic, S., & Palaniswami, M. (2013). Internet of Things (IoT): A vision, architectural elements, and future directions. *Future Generation Computer Systems, 29*(7), 1645–1660.

Gu, T., Pung, H. K., & Zhang, D. Q. (2005). A service-oriented middleware for building context-aware services. *Journal of Network and computer applications, 28*(1), 1–18.

Guo, B., Sun, L., & Zhang, D. (2010, March). The architecture design of a cross-domain context management system. In *Pervasive Computing and Communications Workshops (PERCOM Workshops), 2010 8th IEEE International Conference*, (pp. 499–504). IEEE.

Gluhak, A., Presser, M., Shelby, Z., Scotton, P., Schott, W., & Chevillat, P. (2006). e-Sense reference model for sensor networks in b3 g mobile communication systems. 15th IST Summit 2006, 4–8.

Hwang, G. J. (2006, June). Criteria and strategies of ubiquitous learning. In *IEEE International Conference on Sensor Networks, Ubiquitous, and Trustworthy Computing (SUTC'06)* (Vol. 2, pp. 72–77). IEEE.

Hwang, G. J., & Tsai, C. C. (2011). Research trends in mobile and ubiquitous learning: A review of publications in selected journals from 2001 to 2010. *British Journal of Educational Technology, 42*(4), E65–E70.

Ihantola, P., Ahoniemi, T., Karavirta, V., & Seppälä, O. (2010, October). Review of recent systems for automatic assessment of programming assignments. In *Proceedings of the 10th Koli Calling International Conference on Computing Education Research*, (pp. 86–93). ACM.

Johnson, W. L., Rickel, J. W., & Lester, J. C. (2000). Animated pedagogical agents: Face-to-face interaction in interactive learning environments. *International Journal of Artificial intelligence in education, 11*(1), 47–78.

Kabir, M. A., Han, J., Yu, J., & Colman, A. (2012, June). SCIMS: A social context information management system for socially-aware applications. In *International Conference on Advanced Information Systems Engineering*, (pp. 301–317). Berlin Heidelberg: Springer.

Khan, A. J., Jayarajah, K., Han, D., Misra, A., Balan, R., & Seshan, S. (2013, June). CAMEO: A middleware for mobile advertisement delivery. In *Proceeding of the 11th Annual International Conference on Mobile Systems, Applications, and Services*, (pp. 125–138). ACM.

Kidd, C. D., Orr, R., Abowd, G. D., Atkeson, C. G., Essa, I. A., MacIntyre, B., & Newstetter, W. (1999, October). The aware home: A living laboratory for ubiquitous computing research. In *International Workshop on Cooperative Buildings*, (pp. 191–198). Berlin Heidelberg: Springer.

Kim, J., Lee, D., & Chung, K. Y. (2014). Item recommendation based on context-aware model for personalized u-Healthcare service. *Multimedia Tools and Applications, 71*(2), 855–872.

Klašnja-Milićević, A., Vesin, B., Ivanović, M., & Budimac, Z. (2011). E-Learning personalization based on hybrid recommendation strategy and learning style identification. *Computers & Education, 56*(3), 885–899.

Liang, G., & Cao, J. (2015). Social context-aware middleware: A survey. *Pervasive and Mobile Computing, 17,* 207–219.

Lipponen, L. (2002, January). Exploring foundations for computer-supported collaborative learning. In *Proceedings of the Conference on Computer Support for Collaborative Learning: Foundations for a CSCL Community,* (pp. 72–81). International Society of the Learning Sciences.

Lutolf, R. (1992, November). Smart home concept and the integration of energy meters into a home based system. In *Metering Apparatus and Tariffs for Electricity Supply, 1992. Seventh International Conference,* (pp. 277–278). IET.

Mihailidis, A., Carmichael, B., & Boger, J. (2004). The use of computer vision in an intelligent environment to support aging-in-place, safety, and independence in the home. *IEEE Transactions on Information Technology in Biomedicine, 8*(3), 238–247.

Munoz, L. S., Palazzo, J., & Oliveira, M. (2004). Applying Semantic Web technologies to improve personalisation and achieve interoperability between educational adaptive hypermedia systems. In *Proceedings of the SW-EL workshop at International Conference on Adaptive Hypermedia and Adaptive Web-Based Systems (AH04).* The Netherlands: Eindhoven.

O'Malley, C. (Ed.). (2012). Computer supported collaborative learning (Vol. 128). Springer Science & Business Media.

Perera, C., Zaslavsky, A., Christen, P., & Georgakopoulos, D. (2014). Context aware computing for the internet of things: A survey. *IEEE Communications Surveys & Tutorials, 16*(1), 414–454.

Petite, T. D., & Huff, R. M. (2004). U.S. Patent No. 6,836,737. Washington, DC: U.S. Patent and Trademark Office.

Reddy, M., Pratt, W., Dourish, P., & Shabot, M. M. (2002). Sociotechnical requirements analysis for clinical systems. *Methods of Information in Medicine, 42*(4), 437–444.

Renard, J. M., Bricon-Souf, N., Geib, J. M., & Beuscart, R. (1999). A simulation of dynamic tasks routing to improve cooperation in intensive care units. Studies in health technology and informatics, 31–36.

Rosenberg, M. J. (2001). *E-learning: Strategies for delivering knowledge in the digital age* (Vol. 3). New York: McGraw-Hill.

Satpathy, L. (2006). Smart housing: Technology to aid aging in place: New opportunities and challenges. Mississippi State University.

Schlager, D., & Baringer, W. B. (1997). U.S. Patent No. 5,650,770. Washington, DC: U.S. Patent and Trademark Office.

Simmons, L. A., Dinan, M. A., Robinson, T. J., & Snyderman, R. (2012). Personalized medicine is more than genomic medicine: Confusion over terminology impedes progress towards personalized healthcare. *Personalized Medicine, 9*(1), 85–91.

Soliman, M., Abiodun, T., Hamouda, T., Zhou, J., & Lung, C. H. (2013, December). Smart home: Integrating internet of things with web services and cloud computing. In *Cloud Computing Technology and Science (CloudCom), 2013 IEEE 5th International Conference,* (Vol. 2, pp. 317–320). IEEE.

Sousa, J. P., & Garlan, D. (2002). Aura: An architectural framework for user mobility in ubiquitous computing environments. In *Software Architecture,* (pp. 29–43). US: Springer.

Spencer, R., & Logan, P. (2002). Role-based communication patterns within an emergency department setting. HIC 2002. Proceedings: Improving Quality by Lowering Barriers, 166.

Stone, A. (2014). Online assessment: What influences students to engage with feedback? *The Clinical Teacher, 11*(4), 284–289.

Temdee, P. (2014). Ubiquitous learning environment: Smart learning platform with multi-agent architecture. *Wireless Personal Communications, 76*(3), 627–641.

Verbert, K., Manouselis, N., Ochoa, X., Wolpers, M., Drachsler, H., Bosnic, I., et al. (2012). Context-aware recommender systems for learning: A survey and future challenges. *IEEE Transactions on Learning Technologies, 5*(4), 318–335.

Virone, G., Noury, N., & Demongeot, J. (2002). A system for automatic measurement of circadian activity deviations in telemedicine. *IEEE Transactions on Biomedical Engineering, 49*(12), 1463–1469.

Weiser, M. (1991). The computer for the 21st century. *Scientific American, 265*(3), 94–104.

Winters, N. (2007). What is mobile learning. Big issues in mobile learning, 7–11.

Yu, Z., Yu, Z., & Zhou, X. (2012). Socially aware computing. *Chinese Journal of Computers, 35* (1), 16–26.

Zaıane, O. R., & Luo, J. (2001, June). Web usage mining for a better web-based learning environment. In *Proceedings of Conference on Advanced Technology for Education*, (pp. 60–64).

Zhang, D., Zhao, J. L., Zhou, L., & Nunamaker, J. F., Jr. (2004). Can e-learning replace classroom learning? *Communications of the ACM, 47*(5), 75–79.

Index

P. Temdee and R. Prasad, *Context-Aware Communication and Computing: Applications for Smart Environment*, Springer Series in Wireless Technology,
DOI 10.1007/978-3-319-59035-6

MIX
Papier aus verantwortungsvollen Quellen
Paper from responsible sources
FSC® C105338

If you have any concerns about our products,
you can contact us on
ProductSafety@springernature.com

In case Publisher is established outside the EU,
the EU authorized representative is:
Springer Nature Customer Service Center GmbH
Europaplatz 3, 69115 Heidelberg, Germany

Printed by Libri Plureos GmbH
in Hamburg, Germany